NASA Contractor Report 3993

Shuttle/Spacelab Contamination Environment and Effects Handbook

L. E. Bareiss, R. M. Payton,
and H. A. Papazian

Martin Marietta Aerospace Denver Division
Denver, Colorado

Prepared for
George C. Marshall Space Flight Center
under Contract NAS8-35770

National Aeronautics
and Space Administration

Scientific and Technical
Information Branch

1986

FOREWORD

This edition of the Shuttle/Spacelab Contamination Environment and Effects Handbook was prepared by Martin Marietta Denver Aerospace under contract NAS8-35770. The information presented in this Handbook consists of the data available at the time this edition was published. Indications of errors, omissions or additional data appropriate for inclusion are requested. Please address these or other comments to:

Richard M. Payton
Martin Marietta Denver Aerospace, D1741
P.O. Box 179
Denver, Colorado 80201
(303) 971-5798

or

Edgar R. Miller, ES61
Marshall Space Flight Center, AL 35812
(205) 544-7752
FTS 824-7752

Such corrections will be incorporated in the final update of this Handbook, scheduled for late 1986.

TABLE OF CONTENTS

	Page
Figures	v
Tables	viii
Acronyms	x
Symbols	xiii

		Page
1	INTRODUCTION	1-1
1.1	Philosophy of Handbook and Intended Use	1-1
1.2	Scope and Applicability	1-1
1.3	Users Guide	1-4
2	SHUTTLE CONTAMINATION ENVIRONMENT AND EFFECTS	2-1
2.1	Experiment Design, Assembly, and Preintegration Contamination	2-2
2.1.1	Experiment Design	2-2
2.1.2	Experiment Assembly	2-3
2.1.3	Preintegration	2-3
2.1.4	Design, Assembly and Preintegration References	2-4
2.2	STS Ground Facilities Contamination	2-5
2.2.1	Ground Contamination Overview	2-5
2.2.2	STS Ground Contamination Environmental Data	2-11
2.2.3	Additional Ground Contamination Data	2-39
2.2.4	Ground Facility Analytical Tools and Models	2-41
2.2.5	Ground Contamination Preventative Techniques	2-51
2.2.6	Key Ground Contamination Technical Personnel	2-52
2.2.7	Ground Contamination References	2-53
2.3	Flight Contamination	2-57
2.3.1	Flight Introduction	2-57
2.3.2	Flight Particulate Contamination	2-65
2.3.2.1	Available Flight Particulate Data	2-65
2.3.2.2	Additional Flight Particle Data	2-73
2.3.2.3	Flight Particle Analytical Tools	2-75
2.3.2.4	Flight Particle Protection Techniques	2-79
2.3.2.5	Key Flight Particle Technical Personnel	2-81
2.3.2.6	Flight Particle References	2-82
2.3.3	Flight Molecular Contamination	2-84
2.3.3.1	Available Data	2-84
2.3.3.2	Additional Flight Information	2-111
2.3.3.3	Flight Molecular Tools/Models	2-111
2.3.3.4	Contamination Prevention Techniques	2-119
2.3.3.5	Key Technical Personnel	2-123
2.3.3.6	Flight Molecular Contamination References	2-123
3	SURFACE INTERACTIONS	3-1
3.1	Materials Oxidation/Erosion	3-1
3.1.1	Introduction	3-1
3.1.2	Available Oxidation/Erosion Data	3-2
3.1.3	Additional Oxidation/Erosion Data	3-10
3.1.4	Ground Oxidation-Erosion Simulation	3-12
3.1.5	Oxidation/Erosion Models	3-15

TABLE OF CONTENTS (Concluded)

		Page
3.1.6	Oxidation/Erosion Protection	3-25
3.1.7	Key Personnel	3-27
3.1.8	Oxidation References	3-29
3.2	Surface Glow Phenomena	3-32
3.2.1	Introduction	3-32
3.2.2	Available Data	3-32
3.2.3	Additional Glow Data	3-37
3.2.4	Ground Simulation Studies	3-38
3.2.5	Models for the Shuttle Glow	3-40
3.2.6	Key Technical Personnel	3-50
3.2.7	Glow References	3-51
4	OTHER ORBITAL NATURAL ENVIRONMENTS	4-1
4.1	Neutral Species	4-3
4.2	The Plasma	4-9
4.3	Radiation	4-12
4.3.1	Electromagnetic	4-12
4.3.2	Charged Particles	4-17
4.4	Magnetic and Electric Fields	4-28
4.4.1	Magnetic Fields	4-28
4.4.2	Electric Fields	4-31
4.5	Other Environment References	4-33
Appendix A	SPACECRAFT MATERIALS PROPERTIES	A-1

LIST OF ILLUSTRATIONS

Fig. No.	Title	Page
1-1	STS/SL Experiment Environment Evaluation Flow	1-5
1-2	Shuttle Orbiter Configuration and Coordinates	1-7
2.2-1	KSC Horizontal Payload Ground Flow	2-6
2.2-2	KSC Vertical, Mixed, Life Sciences, and GAS Ground Flows	2-7
2.2-3	VLS Payload Processing Flow	2-8
2.2-4	Federal Standard 209B Particle Size Distributions	2-13
2.2-5	Effect of Federal Standard 209B Size Distributions	2-14
2.2-6	MIL-STD-1246A Surface Cleanliness Levels	2-15
2.2-7	PSA O and C Building Particle Fallout	2-19
2.2-8	STS-41D Canister Particle Fallout Measurements	2-22
2.2-9	OPF Floor Plan and Sampling Locations	2-24
2.2-10	OPF Particle Fallout Measurements	2-25
2.2-11	PCR Contamination Sampling Locations	2-26
2.2-12	Average PCR Fallout Rates for Typical STS Flights	2-27
2.2-13	STS-1, STS-2, STS-3, and STS-4 Ferry Flight Particle Fallout Data	2-29
2.2-14	VPF Average Particle Fallout Distributions	2-31
2.2-15	VPF Particle Test Locations	2-32
2.2-16	SPIF Particle Fallout Data	2-31
2.2-17	Hangar AE Cleanroom Particle Fallout Distributions	2-33
2.2-18	Average PCR NVR Levels	2-37
2.2-19	Particle Shape as a Function of Particle Size	2-44
2.2-20	STS-41B PCR Particle Fallout Data and Linear Approximation	2-44
2.2-21	Obscuration Ratio vs Number of Particles per ft^2	2-46
2.2-22	Hamberg Model of Particle Fallout Rates	2-49
2.3-1	STS-1, STS-2, STS-3, STS-4, and STS-9 POSA and PSA Flight Particle Fallout	2-67
2.3-2	Percentage of Camera-Photometer Frames Showing Particles for STS-2, STS-3, and STS-4	2-69
2.3-3	Percentage pf STS-9 Camera-Photometer Frames Showing Particles	2-69
2.3-4	Reflectance Change Induced by 3 and 20 micron Al Spherical Contamination	2-71
2.3-5	Reflectance Change Induced by Fly Ash Contamination	2-72
2.3-6	Solar Absorptance Change vs Surface Area Obscured	2-72
2.3-7	BRDF Before and After 1176-h. Exposure, Mirror Vertical	2-74
2.3-8	Mirror and Contamination Reflectances	2-80
2.3-9	Pressure During Accent of STS-2	2-85
2.3-10	Comparison of +Y Axis Mass Accumulation During Ascent	2-87
2.3-11	Surface Contamination as Measured by TQCM on Spacelab 1	2-89
2.3-12	Mass Accumulation during STS-3 L2U Engine Firing	2-90
2.3-13	IECM TQCM Surface Contamination on Spacelab 1	2-91
2.3-14	IECM CQCM Surface Contamination Measurements and Sensor Temperature	2-93
2.3-15	IECM Mass Spectrometer, Flight STS-2: Mass Counts vs Time at amu 18	2-95
2.3-16	The Average H_2O Current and Sensor Temperature Measurements Throughout Flight	2-96
2.3-17	IECM Mass Spectrum, STS-3, 7.2 h. MET	2-96

LIST OF ILLUSTRATIONS (Continued)

Fig. No.	Title	Page
2.3-18	Pressure Rise and Composition by IECM Mass Spectrometer in the Payload Bay during the Door Closing Exercise at 167 h. MET on STS-3	2-97
2.3-19	STS-4 Geometry of Contamination Survey Positions	2-98
2.3-20	STS-4 H_2O Counts during Contamination Survey	2-99
2.3-21	STS-4 He Counts during Contamination Survey	2-99
2.3-22	STS-4 Freon 21 (Mass 67) Counts during Contamination Survey	2-100
2.3-23	Signature of a Vernier RCS Firing during Mapping	2-101
2.3-24	Mass Spectrometer Temperature vs MET	2-102
2.3-25	Mass Spectrometer on Free Flying SPAS-01 Subsatellite on STS-7	2-103
2.3-26	Mass Spectrum of Payload Bay on STS-11	2-104
2.3-27	Mass Spectra of the Payload Bay during Astronaut Activity in the Bay	2-106
2.3-28	IECM Reentry Humidity Monitor and Reentry Temperature (Air Sampler) for STS-2, STS-3, and STS-4	2-107
2.3-29	Pressure during Descent of STS-2	2-108
2.3-30	Mass Accumulation during Descent Phase on +Y Axis TQCM Sensors on STS-2 and STS-4	2-110
2.3-31	Skylab Premission Prediction Correlation with Flight Data	2-113
2.3-32	Atmospheric Explorer-D Return Flux Experiment Results	2-114
2.3-33	Illustration of SPACE II Contaminant Transport Function	2-116
2.3-34	Total Direct and Return Flux vs Time in Baseline STS Bay—Standard Atmosphere	2-119
2.3-35	Column Densities vs Time for Baseline STS Bay	2-120
2.3-36	Volume Air Changes Q/V (min^{-1}) vs Vent Area Volume Ratio A/V as a Function of Internal Pressure	2-122
3.1-1	Erosion vs Specimen Film Thickness, STS-5	3-5
3.1-2	Surface Recession (Corrected for Flux Reduction) as a Function of Impingement Angle	3-5
3.1-3	Evaluation of Oxygen Interaction with Materials Experiment III	3-11
3.1-4	Arrhenius Plot of the Reaction Probability of the O + Carbon Reaction	3-13
3.1-5	STS-8 Atomic Oxygen Monitor TQCM 2: Carbon Coated (Facing Out of Bay)	3-16
3.1-6	STS-8 Atomic Oxygen Monitor TQCM 4: Osmium Coated (Facing Out of Bay)	3-17
3.1-7	Kinetics Analysis of TQCM Data of 3.1-6	3-18
3.1-8	Relative Effects of Atomic Oxygen on a Variety of Polymers	3-23
3.1-9	Oxidation Rate of Polyethylene as a Function of Electron Irradiation	3-24
3.1-10	Optical Properties of 96 percent SiO_2 4 percent PTFE Coated Kapton Samples Unexposed and Exposed to LEO Orbital Environment Compared with Uncoated and Unexposed Kapton	3-28
3.2-1	Altitude Variation of Glow Emission	3-33
3.2-2	Spectral Variation of Glow Emission on AE-C	3-34

LIST OF ILLUSTRATIONS (Continued)

Fig. No.	Title	Page
3.2-3	Six Line Average Tracing of Spacecraft Glow	3-35
3.2-4	Percentage of Glow Longer than 6000Å vs Acceleration Voltage of Ions	3-39
3.2-5	Comparison of the OH Nightglow and the Contaminant Glow Spectrum	3-41
3.2-6	The Spectrum of Spacecraft Glow Compared with that Measured in the Laboratory	3-42
3.2-7	A Schematic Representation of the Chemistry Believed to be Responsible for Spacecraft Ram Glow	3-43
3.2-8	Emission from Electron-Irradiated Air at 0.3 mtorr	3-44
3.2-9	Brightness of Shuttle Glow Compared with Atmospheric Explorer Glow	3-47
3.2-10	The Function of the Thruster Glow Intensity on the Engine Pods as a Function of Time After Thruster Firing	3-49
3.2-11	Analysis of Decay of STS-5 from Figure 3.2-10	3-50
3.2-12	Analysis of Decay of STS-3 from Figure 3.2-10	3-51
4-1	Regions of the Earth's Atmosphere	4-1
4-2	General Configuration of the Magnetosphere	4-2
4.1-1	COESA Model Atmosphere	4-4
4.1-2	Temperature vs Altitude	4-5
4.1-3	Temperature above 400 km on Day 356 at 0° Latitude	4-6
4.1-4	Mass Density vs Altitude	4-7
4.1-5	Atomic Oxygen vs Altitude	4-8
4.2-1	Ionic Composition of Solar Minimum Daytime Winter Ionosphere	4-10
4.2-2	Ionospheric Electron Concentration	4-9
4.2-3	Particle Temperature and Velocity	4-11
4.2-4	Daytime Contours of Equal Electron Density in the Orbital Plane of an Alouette Satellite	4-12
4.3-1	Spectral Distribution Curves Related to the Sun	4-13
4.3-2	Solar Spectrum at 225-km Altitude Averaged from Three Scans with a Grazing Incidence Monochromator and a Photomultiplier Type Detector on 23 August 1962	4-15
4.3-2	(cont.)	4-16
4.3-3	X-ray Spectrum Associated with a Large Solar Flare on Feb. 27, 1969	4-17
4.3-4	Energy Spectra of Protons from Several Moderate Size Solar Events Compared with the Galactic Cosmic Ray Spectrum	4-18
4.3-5	Percentage of Interplanetary Fluence Intercepted by Spacecraft in Circular Geocentric Orbits	4-20
4.3-6	Proton Flux Densities at an Altitude of 296 km	4-21
4.3-7	Flux History over 24-h. Period (539-km Circular Orbit at 28.8° Inclination)	4-22
4.3-8	Trapped Electron Omnidirectional Integral Fluxes above Given Energies at 28.5° Inclination	4-23
4.3-9	Trapped Proton Omnidirectional Integral Fluxes above Given Energies at 28.5° Inclination	4-24
4.3-10	Schematic Representation of Adiabatic Charged Particle Motion	4-25
4.3-11	Total Dose Rate Behind a Spherical Shell at 28.5°	4-27

LIST OF ILLUSTRATIONS (Continued)

Fig. No.	Title	Page
4.4-1	Definition and Sign Convention for the Magnetic Elements	4-29
4.4-2	Contours of Constant F (Total Field) for IGRF 1965.0	4-32
4.4-3	Typical Variation of the Ionospheric Conductivities with Height for a Nighttime Ionosphere	4-33

Table No.		
1-1	STS Flight/Data Matrix	1-2
2.2-1	KSC CWA Environment Requirements	2-8
2.2-2	KSC CWA Operations and Maintenance Requirements	2-9
2.2-3	KSC Horizontal Flow CWA Levels	2-9
2.2-4	Other KSC Facility CWA Levels	2-10
2.2-5	VLS Payload Processing Facility Cleanliness Requirements	2-10
2.2-6	Federal Standard 209B Air Cleanliness Classes	2-12
2.2-7	Visibly Clean Levels and Inspection Criteria	2-18
2.2-8	O and C Building Airborne Particle Measurements	2-20
2.2-9	Canister Airborne Particle Measurements	2-21
2.2-10	OPF Airborne Particle Measurements	2-23
2.2-11	PCR Airborne Particle Measurements	2-27
2.2-12	SPIF Airborne Particle Counts	2-30
2.2-13	O and C Building Molecular Environments	2-36
2.2-14	MMSE Canister Molecular Environments	2-36
2.2-15	OPF Molecular Environments	2-36
2.2-16	KSC PCR Molecular Environments	2-37
2.2-17	Orbiter PLB NVR Measurements	2-38
2.2-18	VPF Molecular Environments	2-38
2.2-19	SPIF Molecular Environments	2-39
2.2-20	Orbiter PLB/SL Surface Areas	2-50
2.3-1	Cascade Impactor Launch Particle Densities	2-66
2.3-2	Descent Cascade Impactor Results	2-70
2.3-3	IECM Air Sampler Organics Detected in Most Significant Quantities on STS-2	2-84
2.3-4	Summary of Results of Air Sampler Contaminants during Ascent from STS-2, -3, and -4	2-85
2.3-5	Species from Ascent Air Sampler on STS-9/Spacelab 1	2-86
2.3-6	Mass Accumulation by TQCM during Ascent Phase	2-87
2.3-7	CQCM Net Molecular Mass Accumulation Rates during Ascent	2-86
2.3-8	Accretion Rates Indicated by the CMP during STS-3 Mission	2-92
2.3-9	CQCM Summary: Net Molecular Mass Accumulation Rates	2-92
2.3-10	H_2O Return Flux by IECM Mass Spectrometer and Calculated Column Densities	2-94
2.3-11	IECM Air Sampler Organics Detected in Most Significant Quantities	2-107
2.3-12	Species from Descent Air Sampler on STS-9/Spacelab 1	2-109
2.3-13	Mass Accumulation by TQCM during the Descent Phase	2-108
2.3-14	CQCM Net Molecular Mass Accumulation Rates during Descent Phase	2-110

LIST OF ILLUSTRATIONS (Concluded)

Table No.	Title	Page
3.1-1	Atomic Oxygen Reaction Efficiencies	3-2
3.1-2	Recession and Reaction Efficiency for Organic Films on STS-8	3-3
3.1-3	Oxidation/Erosion "Aging" of Paints	3-3
3.1-4	Oxidation/Erosion of Various Materials	3-4
3.1-5	Oxidation/Erosion of Tensile Strength, STS-8	3-6
3.1-6	Oxidation/Erosion of Elemental Materials	3-7
3.1-7	Strengths of Eroded Ag/Teflon Film	3-8
3.1-8	Tensile Modulus of Exposed Metalized Teflon	3-8
3.1-9	Comparison of Laboratory and Flight Measurements of the Average Probabilities for the Reaction of Atomic Oxygen with Kapton	3-12
3.1-10	Atomic Oxygen Reaction with Polymers	3-14
3.1-11	Materials Protection Test on STS-5	3-26
3.1-12	Mass Loss of Protected and Unprotected Kapton Samples to Low Earth Orbital Environment	3-26
3.2-1	Material Glow Intensities	3-36
3.2-2	Emission for $H + NO \longrightarrow HNO^*$	3-45
3.2-3	Brightness of Glow as Measured on STS-41G and STS-8	3-48
4.3-1	Spectral Distribution of Solar Radiation	4-14
4.3-2	Model Solar Cosmic Ray Spectra	4-18
4.3-3	Gyroradii and Periods of the Motions of Particles in the Guiding Center Approximation	4-28

Acronyms

AE	Atmospheric Explorer
AEDC	Arnold Engineering Development Center
AFGL	Air Force Geophysics Laboratory
AFSD	Air Force Space Division
AIAA	American Institute of Aeronautics and Astronautics
ASTG	Aerospace Test Group
ASTM	American Society for Testing and Materials
ATO	Atomic Oxygen
BRDF	Bidirectional Reflectance Distribution Function
BTS	Bay To Sun
CCAFS	Cape Canaveral Air Force Station
CMP	Contamination Monitor Package
CQCM	Cryogenic Quartz Crystal Microbalance
CTS	Counts
CWA	Controlled Work Area
DC	Direct Current
DE	Dynamics Explorer
DFI	Development Flight Instrumentation
DFRC	Dryden Flight Research Center
DOD	Department of Defense
DMSP	Defense Meteorological Satellite Program
DSP	Defense Satellite Program
DSTF	Delta Spin Test Facility
EAFB	Edwards Air Force Base
ELS	Eastern Launch Site
EOIM	Effects of Oxygen Interaction with Materials
ERBS	Earth Radiation Budget Satellite
ESA	Explosive Safe Area
ET	External Tank
EUV	Extreme Ultraviolet
FOV	Field-of-View
FTIR	Fourier Transform Infrared Spectography
GAS	Get Away Special
GC/MS	Gas Chromatography/Mass Spectroscopy
GEO	Geosynchronous Orbit
IECM	Induced Environment Contamination Monitor
IOCM	Interim Operational Contamination Monitor
IRT	Infrared Telescope
ISO	Imaging Spectrometric Observatory
HEPA	High Efficiency Particle
HOSC	Huntsville Operations Support Center
HVAC	Heating, Ventilation and Air Conditioning
IMU	Inertial Measurement Unit
IR	Infrared
IUS	Inertial Upper Stage
JPL	Jet Propulsion Laboratory
JSC	Johnson Space Center
KSC	Kennedy Space Center
LDEF	Long Duration Exposure Facility
LEO	Low Earth Orbit
LLL	Low Light Level

Acronyms (cont.)

LLLTV	Low Light Level T.V.
LM	Launch Mount
LMSC	Lockheed Missiles and Space Company
LSOC	Lockheed Space Operations Company
LSSF	Life Sciences Support Facility
MAPM	Martin Ascent Particle Monitor
MCD	Molecular Column Density
MDAC	McDonnell Douglas Astronautics Company
MDTSCO	McDonnell Douglas Technical Services Company
MET	Mission Elapsed Time
MFE	Microabrasion Foil Experiment
MKS	Meter-kilogram-second
MMSE	Multi-Mission Support Equipment
MMU	Manned Maneuvering Unit
MSFC	Marshall Space Flight Center
MSIS	Mass Spectrometer Incoherent Scatter
MTF	Mass Transport Factor
NASA	National Aeronautics and Space Administration
NBS	National Bureau of Standards
NCD	Number Column Density
NOAA	National Oceanographic and Atmospheric Administration
NOMAD	Newtonian Orbital Mechanics and Drag
NTS	Nose To Sun
NVAFB	North Vandenberg Air Force Base
NVR	Nonvolatile Residue
O and C	Operations and Checkout
OEM	Optical Effects Module
OMCF	Orbiter Maintenance and Checkout Facility
OMS	Orbital Maneuvering System
OPF	Orbiter Processing Facility
OR	Obscuration Ratio
OSS	Office of Space Sciences
OSTA	Office of Space and Terrestrial Applications
PACS	Particle Analysis Camera for Shuttle
PCR	Payload Changeout Room
PDP	Plasma Diagnostics Package
PGHM	Payload Ground Handling Mechanism
PL	Payload
PLB	Payload Bay
POSA	Passive Optical Sample Assembly
PPF	Payload Processing Facility
PPR	Payload Preparation Room
PRB	Program Review Board
PRCS	Primary Reaction Control System
PSA	Passive Sample Array
PSI	Physical Sciences, Inc.
PSU	Power Supply Unit
PTC	Passive Thermal Control
QCM	Quartz Crystal Microbalance
RF	Return Flux
RMS	Remote Manipulator System

Acronyms (cont.)

RSS	Rotating Service Structure
SCATHA	Satellite Charging At High Altitude
SDIO	Strategic Defense Initiative Office
SE	Support Equipment
SEM	Scanning Electron Microscopy
SEMS	Shuttle Environment Monitoring System
SIA	Shuttle Induced Atmosphere
SIMS	Secondary Ion Mass Spectrometry
SIRTF	Shuttle Infrared Telescope Facility
SL	Spacelab
SMAB	Solid Motor Assembly Building
SMM	Solar Maximum Mission
SPACE	Shuttle/Payload Contamination Evaluation
SPAS	Shuttle Pallet Satellite
SPC	Shuttle Processing Contractor
SPIF	Shuttle Payload Integration Facility
SRB	Solid Rocket Booster
STP	Standard Temperature and Pressure
STS	Space Transportation System
TDRSS	Tracking and Data Relay Satellite System
TIC	Total Ion Current
TQCM	Temperature-Controlled Quartz Crystal Microbalance
TRASYS	Thermal Radiation Analysis System
TTS	Tail To Sun
T/V	Thermal Vacuum
USAF	United States Air Force
UV	Ultraviolet
VAB	Vertical Assembly Building
VAFB	Vandenberg Air Force Base
VAE	Vehicle Airglow Experiment
VC	Visibly Clean
VF	Viewfactor
VIS	Visible
VLS	Vandenberg Launch Site
VRCS	Vernier Reaction Control System
VRM	Venus Radar Mapper
VPF	Vertical Processing Facility
WSTF	White Sands Test Facility
XLV	X Local Vertical
XPS	X-ray Photoelectron Spectroscopy
XVV	Y Velocity Vector
Y-POP	Y Perpendicular to Orbital Plane
YVV	Y Velocity Vector
ZLV	Z Local Vertical
ZVV	Y Velocity Vector

Symbols

A	Area, Avogadro's number, amps
Å	Angstrom, 10^{-10} meters
amu	Atomic mass unit
A_p	Geomagnetic activity factor
atm	Atmosphere
A.U.	Astronomical unit
B	Magnetic field strength
B_0	Solar brightness
c	Velocity of light
°C	Degrees Celsius
C_D	Discharge coefficient
cm	Centimeter
E	Energy, electric field
ev	Electron volt
F	Flow, frequency, field
°F	Degrees Fahrenheit
F10.7	Solar activity factor
ft	Foot
g	Grams, acceleration due to gravity
G	Giga, 10^9
Hz	Hertz
I	Intensity
J	Current
k	Kilo (1,000), frequency constant, Boltzmann's constant, reaction constant
kcal	Kilocalorie
kev	Thousand ev
km	Kilometer
L	Lunar quiet
m	Meter, milli (10^{-3}), mass
M	Mega, 10^3
M_c	Mass column density
Mev	Million ev
mg	Milligrams
ml	Milliliter
mm	Millimeter
n	Nano, 10^{-9}, density
N	Molecular density
N_c	Molecular column density
ng	Nanograms
p	Momentum
℘	Pressure, reaction probability
P	dP/dt
ppm	Parts per million
Q	Flow
R	Radius, gas constant, Rayleigh, magnetic rigidity
R.E.	Reaction efficiency
S	Scattering, amount of material
Sq	Solar quiet
sr	Steradian
t	Time

Symbols (cont.)

t	Time
T	Temperature, transmittance
v	Perpendicular velocity
v	Parallel velocity
V	Volume, velocity, volts
v/v	volume/volume
W	Weight
w/l	Width to Length
+X	Shuttle axis, forward
+Y	Shuttle axis, right
+Z	Shuttle axis, out of PLB
α	Velocity coefficient, adhesion fraction
α_s	Solar absorptivity
γ	Specific heat ratio, C_p/C_v, cone half angle
Δ	Change
ϵ	Emissivity
θ	Angle, fraction of surface covered, latitude
λ	Mean free path
μ	10^{-6} (micro), Micron (micro-meter), incident molecular flux
μm	Micro-meter
ν	Evaporation rate constant
ρ	Reflectivity, density
σ	Scattering cross section, conductivity
ϕ_d	Direct flux
ϕ_r	Return flux
ϕ_{ss}	Self scattered flux
>	Greater than
∇	Gradient

Section 1—Introduction

SECTION 1

INTRODUCTION

1.1 PHILOSOPHY OF HANDBOOK AND INTENDED USE

The information in this Handbook is intended to assist users of the Spacelab (SL) scientific platform on the Space Transportation System (STS). It will assist experimenters to incorporate into their experimental design and procedures features that will minimize the impact of contamination on the performance of their own and other experiments. The contamination information is presented in sufficient detail for use by most experiments. Extensive references, bibliographies, and contacts are provided for use by those experimenters who need more detailed information.

1.2 SCOPE AND APPLICABILITY

The term contamination, as used in this Handbook, refers to both molecular and particulate matter, either deposited on or within the field-of-view (FOV) of a surface and at a level sensible by the instrument or surface. Contamination during all phases of hardware life is addressed. These phases are preintegration (design, manufacture, assembly, test and shipment to the STS launch site), integration (STS launch site activities prior to launch), ascent (the period from STS engine ignition to payload bay (PLB) door opening), orbit (the period during which the PLB doors are open), descent (from PLB door closure to landing site touchdown), postlanding (ground operations prior to PLB opening, including ferry flight), and de-integration (removal from the PLB and post-flight checkout). Measured and anticipated contamination levels, contamination models, contamination effects and protection methods are presented. The available data for facilities at both Kennedy Space Center (KSC) and Vandenberg Air Force Base (VAFB) are presented. This material will be updated as additional data is obtained.

Table 1-1 shows a summary of STS flights up to the time of the publication of this Handbook. As shown in the table, there have been a number of recent flights for which no contamination data are available. Some data may become available from these flights in the future, but the nature, extent, and schedule for these data are unknown.

In addition to contamination, the effects of the space environments at STS altitudes on spacecraft materials are included in the Handbook. The environments included are atmosphere-induced mass loss (most likely due to atomic oxygen), glow of spacecraft surfaces, other atmospheric influences (density, species, temperature, drag), plasma, electromagnetic and particulate radiation, and magnetic and electric fields.

Table 1-1 STS Flight/Data Matrix

Flight Sequence No.	STS Mission No.	Launch Date	Duration, d:h:m	Inclination/ Beta Angle	Altitude, km (nmi)	Contamination Data	Mission	Orbiter
1	STS-1	4-12-81	2:06:20	40.3/ -29 to -19	240-278 (130-150)	POSA, Ground	DFI	Columbia
2	STS-2	11-12-81	2:06:13	38/ -50 to -45	222-259 (120-140)	IECM, Gas Rel Ground	OAST, DFI	Columbia
3	STS-3	3-22-82	8:00:04	38/ -36 to -23	241 (130)	IECM, DFI, PDP Ground (JPL) (Unpub)*	DFI, PLB Door	Columbia
4	STS-4	6-27-82	7:01:09	28.5/ -1 to +20	306 (165)	IECM, CMP, Ground, AFGL	DOD 82-1 PLB Door	Columbia
5	STS-5	11-11-82	5:02:14	28.5/ -25.9 to -7.4	300 (162)	ATO†	SBS/ANIK C-3	Columbia
6	STS-6	4-4-83	5:00:23	28.5/ -21.6 to -18.8	284-291 (153-157)	Ground	TDRS-A	Challenger
7	STS-7	6-18-83	6:02:24	28.5/ +17.5 to +40.9	291-296 (157-160)	SPAS, Ground	SPAS, ANIK PALAPA	Challenger
8	STS-8	8-30-83	6:01:08	28.5/ +36.4 to +29.4	225-300 (120-160)	ATO, Glow Ground, (JPL)	RMS, Earth Obs, INSAT	Challenger
9	STS-9 (SL-1)	11-28-83	10:07:47	57/	250 (150)	IECM, UV Glow Ground	SL-1	Columbia
10	STS-41B	2-3-84	7:23:16	28.5/ -26.8 to +4.5	326 (176)	Ground (JPL ATO)	SPAS, PALAPA/ WESTAR, MMU	Challenger
11	STS-41C	4-6-84	6:23:40	28.5/ -12.0 to +12.1	504 (272)	Ground	SMM, LDEF	Challenger
12	STS-41D	8-30-84	6:00:56	28.5/ -2.0 to +17.7	332 (179)	Glow, Ground	WESTAR, SBS LEASAT, OAST	Discovery
13	STS-41G	10-5-84	8:05:23	57/ -53.8 to -22.2	352 (190)	JPL-ATO (Ground)	ERBS	Challenger
14	STS-51A	11-8-84	7:23:44	28.5/ -27.8 to -1.3	361 (195)	Ground	ANIK, PALAPA/ WESTAR Rec.	Discovery
15	STS-51C	1-24-85	3:01:33	28.5/ N/A	N/A	Ground, (IOCM?)	DOD/IUS	Discovery
16	STS-51D	4-12-85	6:23:55	28.5/ -4.4 to +15.9	465 (251)	Ground	LEASAT/Garn	Discovery
17	STS-51B (SL-3)	4-29-85	7:00:08	57/ -21.9 to -38.1	342 (190)	Ground	SL-3	Challenger

Table 1-1 (concl)

Flight Sequence No.	STS Mission No.	Launch Date	Duration, d:h:m	Inclination/ Beta Angle	Altitude, km (nmi)	Contamination Data	Mission	Orbiter
18	STS-51G	6-17-85	7:01:39	28.5/ 18.0 to 44.1	342 (190)	Ground	Arabsat/ Telesat/SDI	Discovery
19	STS-51F (SL-2)	7-29-85	7:22:45	49.5/ +32.0 to -4.5	351 (195)	Glow, IRT Contam (Cloud)	SL-2 Astrophysics	Challenger
20	STS-51I	8-27-85	7:02:27	28.5/ 5.9 to 28.9	392 (218)	Ground	LEASAT/ RCA/ASC-1	Discovery
21	STS-51J	10-3-85	4:01:44	28.5/	515 (278)	Ground	DOD	Atlantis
22	STS-61A	10-30-85	7:00:44	57.0/ -24.9 to -57.8	333 (180)	(Ground)	GLOMR	Challenger
23	STS-61B	11-26-85	6:21:04	28.45/ -9.4 to -35.4	387 (209)	(Ground)	MORELOS AUSSAT SATCOM	Atlantis
24	STS-61C	1-12-86	5:01:04			(Ground, PACS)	RCA Ku Hitchhiker	Columbia
25	STS-51L	1-28-86	Vehicle Lost					Challenger

*Terms in Parentheses Indicate Unpublished Information
†ATO Indicates Atomic Oxygen

Note:
Acronyms are defined in the acronym list at the front of the handbook.

1.3 USER'S GUIDE

Wherever practical, the information presented in the Handbook has been organized into data, model, protective measures and reference sections. This will help users to find specific information concerning an environment of interest.

Examples of the use of data, models, and effects information are provided wherever appropriate.

Figure 1-1 shows an overall evaluation flow for the environments discussed in this Handbook. The initial step of the flow (regarding STS/SL contamination control requirements) deals with interface requirements such as outgassing limits for nonmetallic materials, external cleanliness, orbital venting limitations, and other environmental interface requirements. These types of requirements are not covered in detail in this context, since such requirements are dealt with in the experiment interface documentation. Rather, the Handbook is intended for use by an experimenter to determine the effects, if any, of contamination upon his hardware

For contamination evaluation, the flow depicts three general areas of concern: changes in the optical properties (solar absorptance, emittance, reflectance, transmittance, etc.); molecules or particles within the FOV of an instrument while on orbit; and mechanical or electrical interference in the operation of experiment hardware. For some experiments, such as the Isotopic Stack-Measurement of Heavy Cosmic Ray Isotopes of SL-1, which consist of a passive stack of visual track detectors, a brief examination of the data herein may indicate that the STS/SL contamination environment is not a concern. For other experiments, the initial evaluation may indicate a potential for contamination susceptibility. The suggested course in this case is parallel for each area of contamination effects. The degrees of degradation of performance tolerable by the experiment, and the levels of contamination and corresponding effects, using the information from this Handbook, should be predicted. The allowable degradations may then be compared to the predicted levels and effects. If an incompatibility appears, the experiment designer may investigate protective measures for his hardware, also presented herein.

An important contamination concern which cannot be addressed herein involves mission compatibility of payload and experiments to avoid cross contamination, and compatibility of operational timelining to avoid undesirable exposure during periods of higher contamination levels (e.g., first few hours on orbit, dumps and vents, engine firings, high payload bay temperatures). If the above evaluation demonstrates a sensitivity to contamination, experimenters are urged to bring this to the attention of the mission management so that potential deleterious effects can be minimized.

In the areas of materials oxidation, glow, and other natural environments encountered while on orbit, a similar approach is suggested. Less detailed analysis flows for these environments are shown in Figure 1-1, in keeping with the limited available data for oxidation and glow

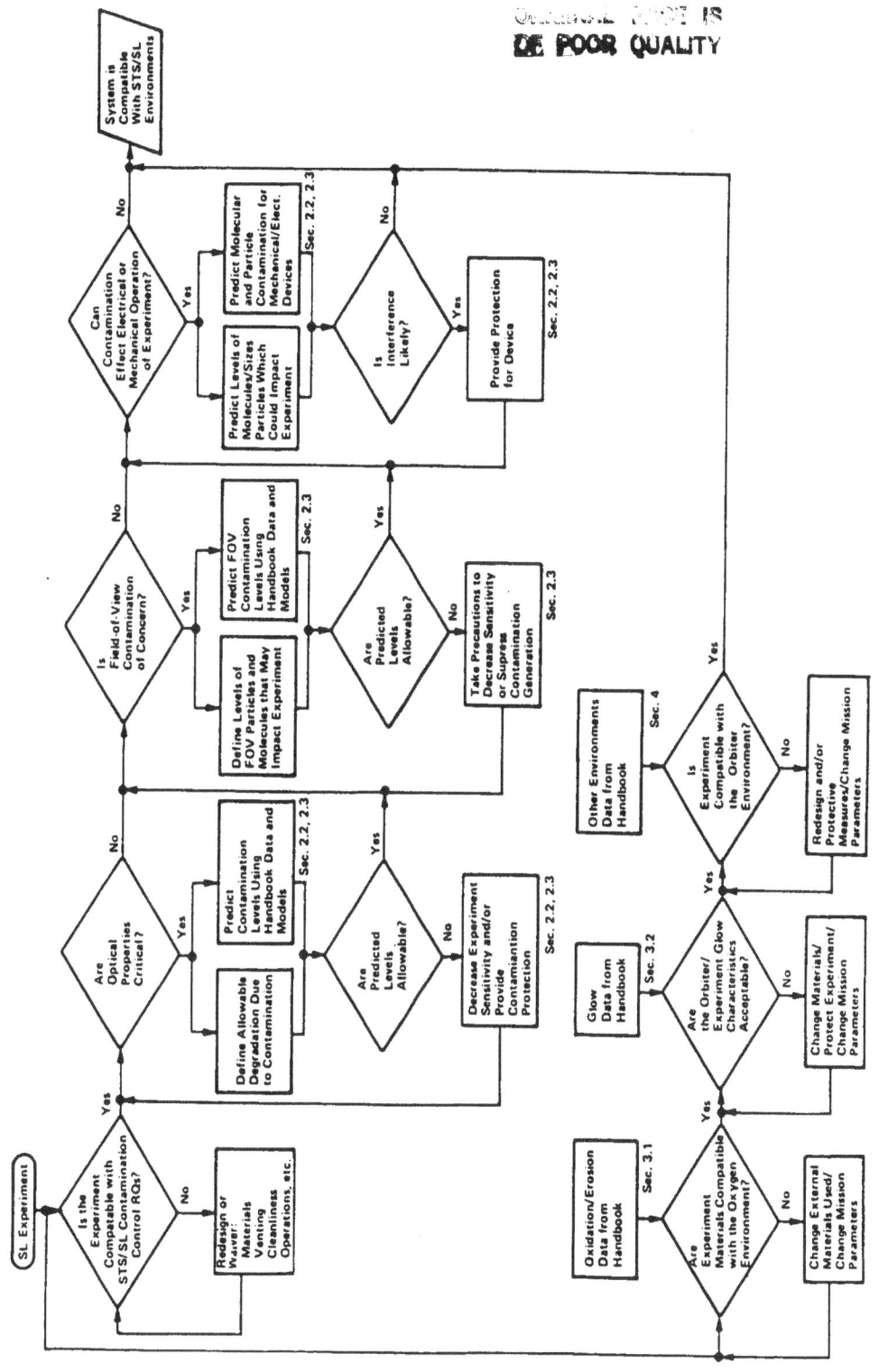

Figure 1-1 STS/SL Experiment Environmental Evaluation Flow

effects, and the limited depth of presentation of other environments within the scope of this Handbook.

Figure 1-2 shows the Space Shuttle Orbiter with a standard coordinate system. As noted in the figure, coordinates used by the Induced Environment Contamination Monitor (IECM) differ, in that $X_O = -X_0$ and $Z = -Z_0$.

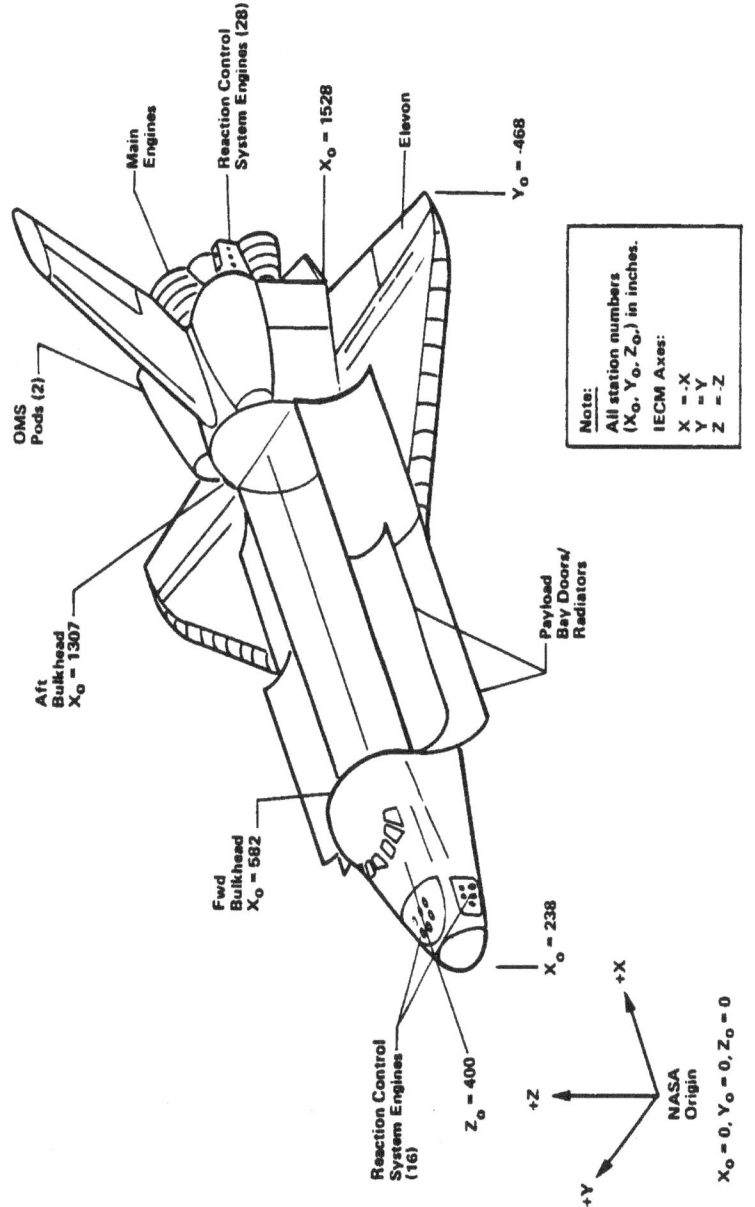

Figure 1-2 Shuttle Orbiter Configuration and Coordinates

Section 2—Contamination Environment and Effects

SECTION 2

SHUTTLE CONTAMINATION ENVIRONMENT AND EFFECTS

As stated previously, the term contamination refers to molecular and particulate matter, both deposited on and within the field-of-view of surfaces.

Particulate contamination includes solid and liquid particles. Settling of particles suspended in the air of ground facilities is traditionally the primary concern for particulate contamination. Particles may also migrate from one surface to another, however, during the dynamic environments encountered in STS launch and orbital operations. The crews of STS flights have often commented on the quantities of particles leaving the STS PLB during Shuttle flights.

The effects of particles on surfaces include obscuration of optical surfaces, aberration of optical images, and changes in absorptance, emittance, and specular and spectral properties. Experimenters have some control over exposure to airborne particles while their hardware is on the ground, and may limit the number of particles deposited on a surface through the use of covers, cleaning operations, selection of facilities, and limitations on exposure times. Control over particle contamination during STS launch and orbital operations is more difficult to enact, but include deployable covers, surface orientation selection, experiment design, and cleanliness of adjacent surfaces.

Particles within the field-of-view of experiments can interfere with photographic images and optics used for star fixes, and can increase the PLB background radiation environment by scattering and emitting light into the bay. Individual experiments may have little or no control over particles within their field-of-view. Protective measures include imposition of stringent cleanliness requirements upon other PLB surfaces and scheduling observation times around likely particle generating events.

The initial goal of any experimenter should be to determine the degree of contamination control necessary. This is found by determining ultimate sensitivity of the experiment for proper on orbit operation. Once determined, the experimenter can then establish the necessary controls. Sensitivities can be compared with data herein to determine if the experiment will be highly sensitive, moderately sensitive, or insensitive to the STS/SL environment. Depending on the results additional protective measures may be required, or no unique control may be necessary.

Molecular contamination is also a concern from the standpoint of both field-of-view interference and deposition effects. Molecular deposition is likely to occur both on the ground and on orbit. Molecules released by paints, solvents, nonmetallic materials, and other sources can be carried by facility air to sensitive surface on the ground. Cleaning materials can also be a source of ground molecular contamination. Orbital deposition can occur through direct transport from sources (direct flux), or through scattering. Scattering by surfaces is termed reflection, scattering by contaminant molecules is termed self-scattering, and scattering by ambient molecules is termed return flux.

Effects of molecular deposits include changes in surface optical properties (solar absorptance, transmittance, reflectance, diffraction through droplet formation, etc.). Field-of-view molecular effects include electromagnetic emission, absorption and scattering.

2.1 EXPERIMENT DESIGN, ASSEMBLY AND PREINTEGRATION CONTAMINATION

Ground activities establish the background contamination levels for space hardware. The design of hardware influences the hardware's susceptibility to contamination and its cleanability. The design and the selection of materials determine the hardware's propensity for self-contamination or cross-contamination of other payloads. The facilities and activities the hardware endures contribute to the cleanliness level present at the time of launch. Cleaning activities while on the ground enhance the cleanliness of hardware, but are never completely successful in removing all contaminants. Design, material selection, protection techniques and selection of facilities for ground activities are all important, therefore, in ensuring performance on orbit.

Each individual piece of experiment hardware will follow a unique path through manufacturing, assembly and test facilities. These facilities may be both NASA and experimenter controlled. Guidelines therefore, rather than data for specific preintegration facilities are presented herein. It is critical that experimenters develop an approach to contamination control that spans the lifetime of their hardware. The suggested approach is through the development of a contamination control plan to ensure contamination control during all hardware phases. The penalties for initiating contamination control or discovering contamination problems upon arrival at the launch site can involve significant time and cost impacts.

2.1.1 Experiment Design

Required cleanliness levels for desired performance are important inputs to the design of hardware for use in space. These include cleanliness of fluid systems, mechanical bearing surfaces, and optical surfaces such as lens systems or thermal control surfaces. Should the hardware require specific cleanliness levels for proper operation, consideration should be given to design options which minimize the effects of contamination. In the event of no functionally defined requirements, the hardware still must be visibly clean (VC), as defined in Reference B-1, in order to be installed and launched in the Space Shuttle PLB. All hardware should, therefore, be designed to facilitate cleaning while at the launch site. The design should also preclude areas in which contaminants could be trapped, since trapped contaminants may be agitated during STS launch or flight, and could migrate to sensitive surfaces on the payload or on other experiments.

A general approach to identifying contamination allowables is to begin with acceptable performance degradation levels. Contamination effects data and analysis tools are then used to convert the performance allowables to allowable contamination levels. These can then be compared to predicted or expected contamination environments to evaluate compatibility. Incompatibilities can be dealt with through protection techniques or design modifications.

Nonmetallic materials used in orbit on SL and STS must meet the vacuum outgassing requirements of Reference B-2. In general, this means that materials must outgas less than 1 percent of their total weight during 24 hours of vacuum exposure at 125°C, and less than 0.1 percent of the total mass can be collected on a 25°C surface in the standard test described in the reference. If materials needed for a specific application do not meet the test criteria, waivers can be considered based on quantities used and the proximity of sensitive hardware. The exclusive use of qualified materials, however, still does not ensure sensitive surfaces will remain uncontaminated. Use of a low outgassing paint on a sunshade for a highly sensitive optical system, for example, may still deposit unwanted materials on the optics if small quantities of condensibles are outgassed. The modeling techniques described in Section 2.3.3 should be employed to verify that materials are compatible with their applications. References B-1 and B-3 provide further guidelines for these and other aspects of contamination control during space hardware design.

2.1.2 Experiment Assembly

In view of the less than 100 percent efficiency of cleaning operations, and the difficulty of designing hardware which provide complete access for cleaning and an absence of contamination traps, careful consideration should be given to the cleanliness of the facilities in which components are manufactured, assembled and tested. In general, some form of controlled environment, including high-efficiency particulate air (HEPA) filters and personnel trained in contamination control is recommended for all but the most easily cleaned space hardware. During assembly, surfaces should be examined, and cleaned when necessary, prior to becoming inaccessible during assembly. Correlations of air and surface cleanliness will be provided in Section 2.2 and its subsections.

2.1.3 Preintegration

Preintegration includes checkout and testing of flight articles after assembly, and shipment to the launch site. Inadequate contamination control provisions during these activities can counteract the benefits of precautions taken during manufacture and assembly.

Hardware can be particularly vulnerable to contamination during testing. During any transport operations to test facilities, adequate packaging must be provided to protect the equipment. The cleanliness of the test facilities should be verified to be compatible with the desired cleanliness of the hardware being tested. In the case of new designs or state-of-the-art approaches, some sensitive hardware may be the most sensitive "monitor" to enter a facility. In other words, the traditional methods of measuring contamination may not be able to detect contamination levels that could degrade very sensitive hardware. Special care must be taken with such designs.

Guidelines for packaging equipment for contamination control are contained in References B-1 and B-3. The guidelines apply equally to temporary packaging for transit between controlled areas and long distance shipments.

2.1.4 <u>Design, Assembly and Preintegration References</u>

B-1. "Specification Contamination Control Requirements for the Space Shuttle Program", SN-C-0005A, National Aeronautics and Space Administration (NASA) Johnson Space Center (JSC), Houston, Texas, January 1982.

B-2. Leger, L. J., "General Specification Vacuum Stability Requirements of Polymeric Materials for Spacecraft Application", SP-R-0022A, NASA JSC, Houston, Texas, 9 September 1974.

B-3. "Contamination Control Handbook", NASA SP-5076, Sandia Labs for NASA Marshall Space Flight Center (MSFC), 1969.

2.2 STS GROUND FACILITIES CONTAMINATION

2.2.1 Ground Contamination Overview

This section provides information on contamination environments in the STS ground facilities, and includes environmental data for integration facilities, post-landing facilities, ferry flight, and deintegration facilities. Recorded data and available models for application of the data to actual spacecraft are presented. Data on the effects of contamination will be presented in the sections dealing with orbital contamination to minimize duplication.

2.2.1.1 STS/SL Processing Flows and Facilities

Spacelab was designed to use the horizontal flow for integration with the STS Orbiter. The KSC ground flow for horizontal payloads is shown in Figure 2.2-1 (Ref. C-1). The standard SL ground flow begins at the KSC initial receipt location. SL experiments are transported to the Operations and Checkout (O and C) Building for checkout and installation of hardware and experiments on the SL modules or pallets. The entire SL is then transported via the horizontal canister transporter to the Orbiter Processing Facility (OPF) for installation in the Orbiter PLB. After installation, the PLB is sealed and the Orbiter is towed to the Vertical Assembly Building (VAB), where it is rotated to the vertical position and mated with the external tank (ET) and solid rocket boosters (SRB). The PLB remains sealed while in transit and in the VAB. PLB purges are used after the PLB is sealed in the OPF, before the Orbiter is transported to the VAB, and from the time the Orbiter is mated to the ET and SRBs until launch. From the VAB, the STS assembly is transported to the launch complex. At the launch complex, the PLB may be opened in the Payload Changeout Room (PCR) of the Rotating Service Structure (RSS) for final cleaning, checkout, or installation of hardware.

Payloads other than SL, or SL elements not installed on SL in the O and C Building may use other KSC ground flows. Figure 2.2-2 shows ground flows for vertical payload processing, mixed payloads, life science payloads and getaway special (GAS) processing (Ref. C-1). These flows may be encountered by special SL hardware, or by systems that fly on both SL and other STS flights. Additional facilities used in these processing flows are the Solid Motor Assembly Building (SMAB), Explosive Safe Area (ESA) 60A and the Delta Spin Test Facility (DSTF) (all used for upper stage processing), the Vertical Processing Facility (VPF) (used for stacking vertical payload assemblies), various other payload processing facilities (PPF) (Hangars AE, AM, AO and S), the Life Science Support Facility (LSSF), and the Shuttle Payload Integration Facility (SPIF) (Defense Department vertical payload processing).

Some SL missions may be high orbital inclination missions flown from Vandenberg Launch Site (VLS). Figure 2.2-3 shows the flow and facilities for VLS payload processing (Ref. C-2). Horizontal payloads (such as SL) will be installed into the Orbiter at the Orbiter Maintenance and Checkout Facility (OMCF). The Orbiter will then be towed to the Launch

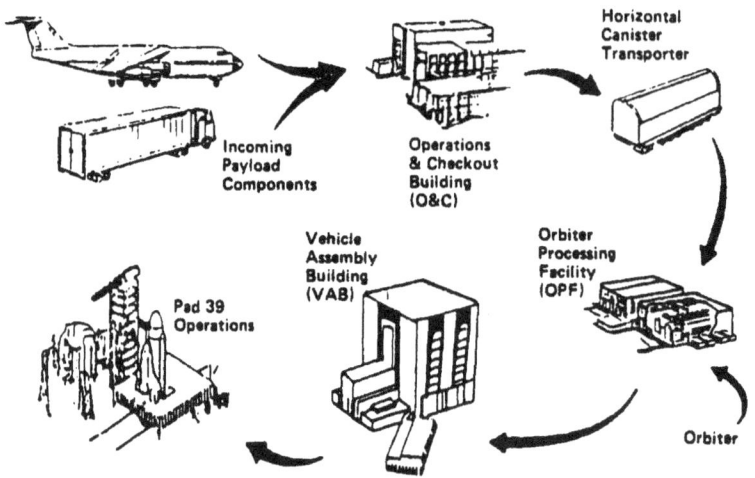

Figure 2.2-1 KSC Horizontal Payload Ground Flow (Ref C-1)

pad area, where it will be erected on the launch mount (LM) and mated with the ET and SRBs. Any vertical payloads will be prepared in the Payload Preparation Room (PPR) and transferred to the Orbiter via the Payload Changeout Room (PCR). Access to the PLB is provided in the PCR.

Primary landing sites for STS missions are at KSC and VAFB. After landing, the Orbiter is towed to the OPF at KSC, and the OMCF at VAFB, where the PLB is opened. SL experiment hardware can then be removed, or the SL may be removed intact. At KSC, the intact SL will be transported to the O and C Building for deintegration. From VAFB, the SL will probably be shipped to the KSC O and C Building.

Two alternate landing sites have already been used for STS landings. These are at Edwards Air Force Base (EAFB)/Dryden Flight Research Center (DFRC) and the White Sands Test Facility (WSTF). For landings at these or other contingency sites, the PLB may be opened to secure or safe the payload. The Orbiter/SL will then be transported to one of the primary landing sites via a Boeing 747 ferry flight.

2.2.1.2 Ground Facility Environment Overview

NASA KSC provides five levels of environmental control in its processing facilities. Table 2.2-1 summarizes the minimum environmental requirements for the five environmental levels, and Table 2.2-2 describes the operations and maintenance requirements (Ref. C-3). Since the facility areas are not clean rooms as defined in Reference C-4, the controlled environments at KSC are termed controlled work areas (CWA). Table 2.2-3 presents the CWA cleanliness levels for the standard horizontal processing flow facilities. Table 2.2-4 presents the levels for the other cargo processing facilities mentioned previously (Ref. C-3 and C-5). In general,

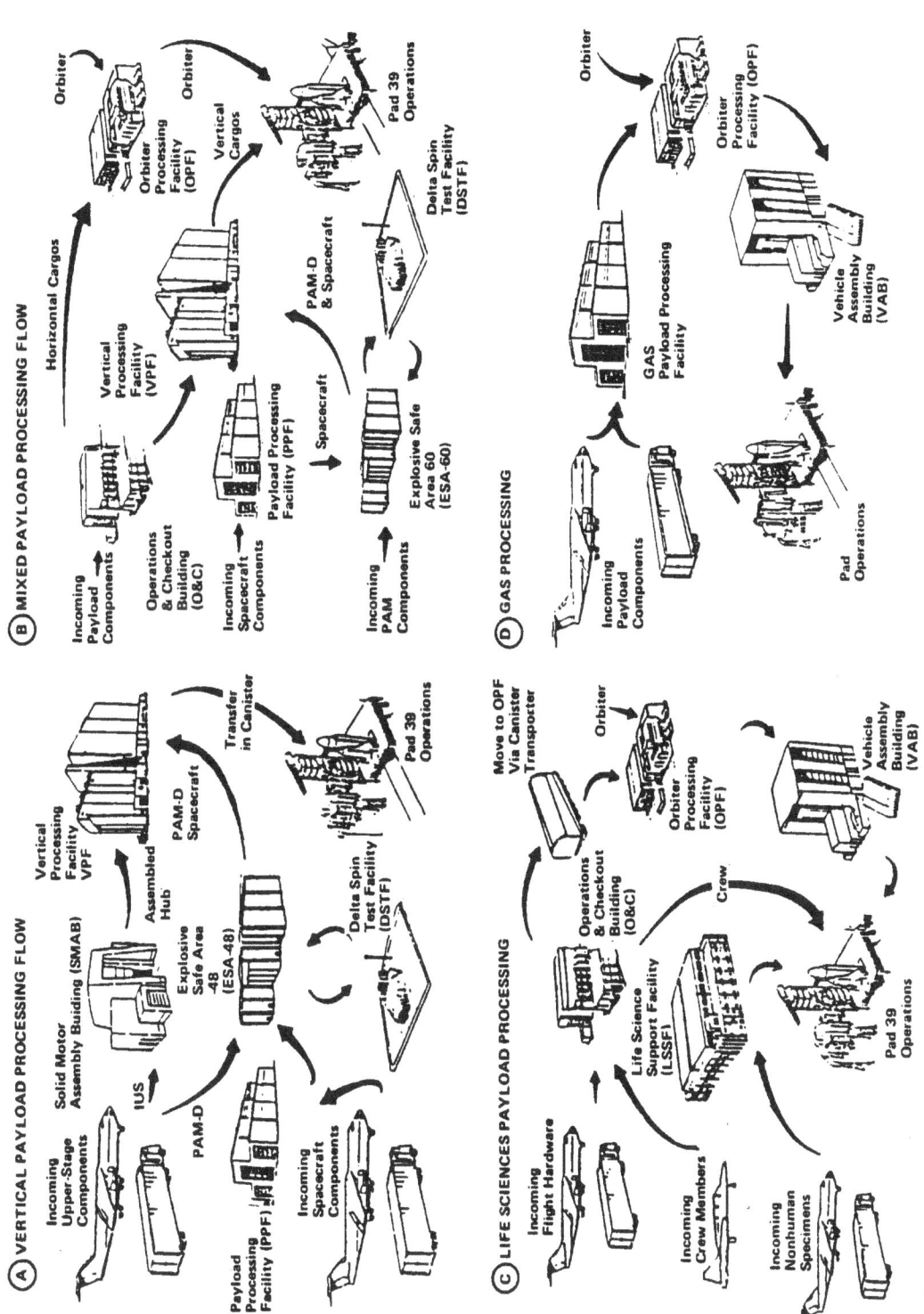

Figure 2.2-2 *KSC Vertical, Mixed, Life Sciences, and GAS Ground Flows (Ref C-1)*

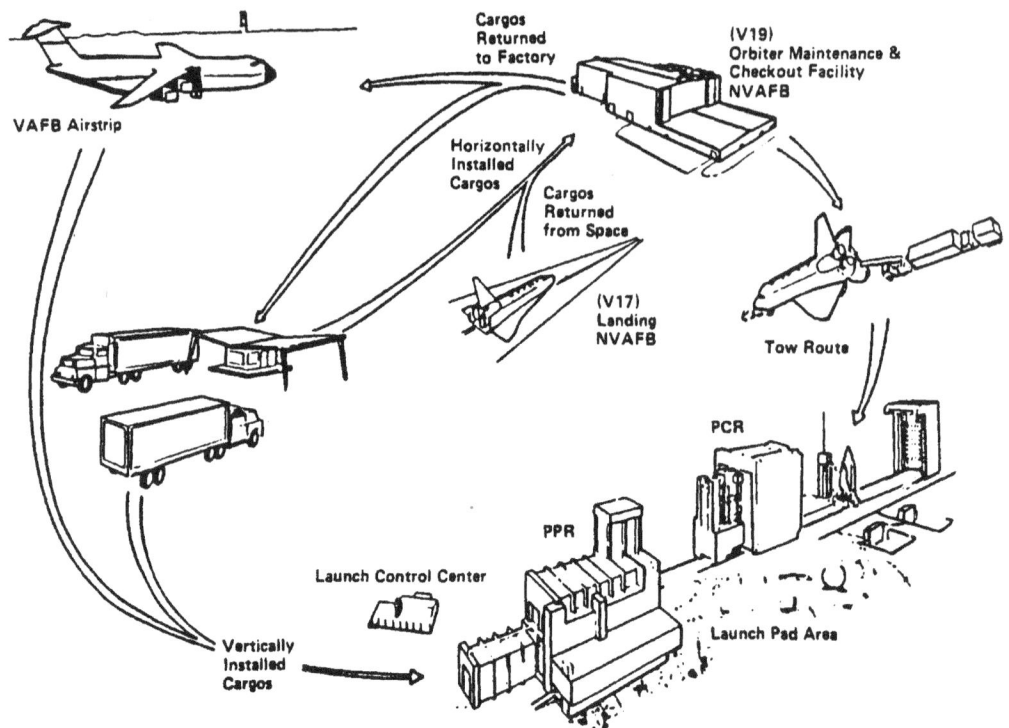

Figure 2.2-3 VLS Payload Processing Flow (Ref C-2)

Table 2.2-1 KSC CWA Environment Requirements (Ref C-3)

Air Flow Type		Level 1 Laminar	Level 2 Laminar	Level 3 Non-laminar	Level 4 Non-laminar	Level 5 Non-laminar
Maxium Airborne Particulate Counts, per ft^3	– Req 0.5 μ – Req 5.0 μ – Monitoring	5,000 30 Continuous	10,000 65 Continuous	50,000 300 Continuous	100,000 700 Continuous	300,000 1,000 Monthly
Temperature, °F	– Requirement – Monitoring	71 ± 6 Continuous	71 ± 6 Continuous	71 ± 6 Continuous	71 ± 6 Continuous	71 ± 6 Monthly
Relative Humidity, %	– Requirement – Monitoring	50 Max Continuous	50 Max Continuous	50 Max Continuous	50 Max Continuous	50 Max Monthly
Maximum Particle Fallout	– Goal* – Monitoring	Level 200 Continuous	Level 200 Continuous	Level 500 Continuous	Level 750 Continuous	Level 1000 Semiannually
Maximum NVR, mg/0.1m^2/month	– Requirement – Monitoring	1 Continuous	1 Continuous	1 Continuous	1 Continuous	2 Annually
Maximum Volatile Hydrocarbons (PPM), v/v	– Requirement – Monitoring	15 Every 2 weeks	15 Every 2 weeks	15 Every 2 weeks	15 Every 2 weeks	N/A N/A
Minimum Positive Pressure	– Requirement – Monitoring	0.05-in. H_2O Daily	0.05-in. H_2O Daily	0.05-in. H_2O Daily	0.02-in. H_2O Daily	N/A N/A
Minimum Air Changes	– Requirement	20/hour	20/hour	6/hour	4/hour	2/hour

*Levels per MIL-STD-1246A for a 24-h period

the environmental requirements of Table 2.2-1 have been met during most STS processing operations. Section 2.2.2 presents the available data to date.

Table 2.2-2 KSC CWA Operations and Maintenance Requirements (Ref C-3)

CWA Levels		CWA Levels 1 & 2	CWA Levels 3 & 4	CWA Level 5
Cleanliness Level		VC	VC	VC
Clean Room Clothing	– Garments	Coveralls	Smocks	Smocks (Opt)
	– Head Covering	Hoods	Caps/Hoods	Caps/Hoods (Opt)
	– Shoe Covering	Booties	Booties (Opt)	Booties (Opt)
	– Gloves	Optional	Optional	Optional
Personnel Access		Limited	Limited	Limited
Operations		Rigidly Controlled	Rigidly Controlled	Controlled
Tools & Fixtures		Rigidly Controlled	Rigidly Controlled	Controlled
Material Entry		Rigidly Controlled	Rigidly Controlled	Controlled
Floors (Including Platform Floors)	– Inspection	Weekly	Weekly	Monthly
	– Vacuum	Daily	Twice Weekly	Weekly
	– Damp Mop	Daily	Twice Weekly	Weekly
Walls & Ledges below 12 ft	– Inspection	Semiannually	Semiannually	Semiannually
	– Cleaning	As Required	As Required	As Required
Ceilings	– Cleaning	As Required	As Required	As Required
Platform Structures	– Inspection	Weekly	Weekly	Weekly
	– Cleaning	Daily	Monthly	As Required
GSE	– Inspection	Biweekly	Bimonthly	Bimonthly
	– Cleaning	Weekly	Monthly	Monthly
Cranes, Hoists, etc	– Inspection	Quarterly	Quarterly	Quarterly
	– Cleaning	As Required	As Required	As Required

Table 2.2-3
KSC Horizontal Flow CWA Levels (Ref C-3)

Facility	Location	CWA Level
O&C	High Bay	5
	Test Stands & Specific Work Areas	4
	Room 3299A	4
	Offline Labs	5
	ATM Cleanroom	3
Canister	N/A	4
OPF	P13-A	4
Pad	PCR	4

Table 2.2-4
Other KSC Facility CWA Levels (Ref C-3, C-5)

Facility	Location	CWA Level
AE	High Bay	2
AO	High Bay	4
AM	High Bay	5
S	High Bay	4
ESA-60A	High Bay	4
DSTF	High Bay	4
VPF	High Bay Air Lock	4 5
SAEF-2	High Bay	4
SPIF	Various	Equiv to 4

The environmental requirements for various areas of the OMCF, PPR and PCR at VAFB are presented in Table 2.2-5 (Ref. C-2). Data on the actual measured environments will be included in Section 2.2.2 as they become available.

Ground facility contamination environments of concern to SL experimenters include volumetric measurements of airborne particles and hydrocarbons, and deposition of particles and molecules on surfaces. Related environments that may also affect contamination deposition are airborne salt, relative humidity and temperature.

Considerable work has been conducted recently to develop models relating volumetric particle measurements to surface deposition and degradation. Comparable models for volumetric and deposited molecules have not yet been developed.

Table 2.2-5
VLS Payload Processing Facility
Cleanliness Requirements (Ref C-2)

Facility	Location	CWA Level
OMCF	– Environmental Enclosure – Trans Air Lock – High Bay Area – Payload Storage – Payload Deservice – Mezzanine – Payload Bay	– Similar to 4 – Similar to 4 – Similar to 5 – Similar to 4 – Similar to 4 – Similar to 4 – Similar to 4
PPR	– Air Lock – Erection Room/ Transfer Tower – Checkout Cells – Storage Rooms – PL SE Rooms	– Similar to 4 – Similar to 4 – Similar to 4 – Similar to 4 – Similar to 4
PCR	– N/A	– Similar to 4

2.2.2 STS Ground Contamination Environmental Data

Considerable quantities of data have been collected in the integration facilities at KSC for most of the Shuttle flights to date. The data have been collected by several agencies, including NASA, its contractors, the Air Force, and the Aerospace Corporation.

Data collected during the first four Shuttle flights raised some concerns about the adequacy of KSC facility environmental control. Rather severe contamination occurring while STS-6 and the Tracking and Data Relay Satellite System-A (TDRSS-A) payload were being processed emphasized these concerns. Steps taken to improve the environmental controls, and the transition from checkout to operational activities have greatly reduced the typical contamination levels found in the KSC integration facilities. More recent data is therefore more representative of the current situation at KSC.

Available data is more limited for deintegration facilities, post-landing facilities, and ferry flights. Deintegration occurs, in general, in the OPF, O and C and OMCF buildings. While any data collected during deintegration in these facilities is not currently available, the environment may be similar to that for integration activities. It is possible, however, that cleanliness discipline occurring during deintegration would not be equivalent to that employed for integration activities. Currently available data for non-KSC post-landing facilities and ferry-flights is limited to that collected by Orbiter PLB contamination monitors.

The following subsections summarize the available ground contamination data for integration facilities. Data for the primary SL facilities (O and C, OPF, and PCR at KSC, OMCF and PCR at VAFB) are presented first, followed by data for other facilities.

2.2.2.1 Ground Facilities Particulate Data

In general, two types of particle data have been collected in the KSC cargo facilities: counts of the number of particles with diameters larger than 0.5 μm suspended in 1 ft^3 of facility air; and the number of particles with diameters larger than either 1 or 5 μm deposited on 1 ft^2 of surface in 24 hours. Two government publications define levels of cleanliness for these types of particle data: Fed. Std. No. 209B (Ref. C-4) and MIL-STD-1246A (Ref. C-6). These documents are the source of the mixed units (meter-kilogram-second (MKS) and British Engineering) traditionally used for measurements of these environments. Throughout this Handbook traditional units will be used when they are necessary for clarity, rather than an exclusive MKS system. When data is presented in a consistent MKS set of units, these will be presented along with a conversion to traditional units.

Fed. Std. No. 209B defines air cleanliness classes as shown in Table 2.2-6. It further elaborates these classes with Figure 2.2-4, and states that other air cleanliness classes may be defined by the intercept point on the 0.5 μm line in Figure 2.2-4 with a line parallel to the three established curves. (It may be noted that the three established curves are not themselves parallel, since the class 10k line passes through 65 at 5 μm, while the class 100k line similarly passes through 700 instead of 650. The resulting error is small, though, and lines parallel to either established curve are commonly used.) Thus, air cleanliness data may be referred to in terms such as class 14.5k if the distribution of airborne particles is such that the recorded data remains below a line parallel to the established curves. This type of data is not always collected, however. Sometimes particle counters simply report the number of particles/ft^3 greater than 0.5 μm. Figure 2.2-5 shows two particle distributions. Both have 14,500 particles/ft^3. As will be discussed in Section 2.2.4, however, they could have significantly different impacts on spacecraft systems.

Table 2.2-6
Federal Standard 209B Air Cleanliness Classes (Ref C-4)

Maximum Number of Particles per Cubic Foot (per Liter) 0.5 micron & Larger	Traditional Units Class (Metric System)	Maximum Number of Particles per Cubic Foot (per Liter) 5.0 microns & Larger
100 (3.5)	100 (3.5)	See note at Bottom of Figure 2.2-4
10,000 (350)	10,000 (350)	65 (2.3)
100,000 (3,500)	100,000 (3,500)	700 (25)

Reference C-3, the KSC Cargo Facilities Contamination Control Plan cites Reference C-4 for airborne particle counting techniques. Two techniques are described. Automated counters using light scattering detection techniques may be used for particles with diameters greater than 0.5 μm. These devices usually operate continuously and are the most commonly used monitors at KSC. The other method described consists of drawing a known quantity of air through a membrane filter and counting the particles collected using microscopic techniques. This method is restricted to particle sizes greater than 5.0 μm, and is also restricted to air classes greater than class 10,000 in order to collect statistically valid numbers of particles. This technique is thus not commonly used. Reference C-4 also allows any other technique which can demonstrate accuracy and repeatability equivalent to that of the previously described methods. Monitoring methods are not typically specified in the data reports included herein. Because of their much more common usage, unless otherwise specified, the data is assumed to have been collected using an automated light scattering detector. Reference C-7 describes the contamination control implementation for the KSC Shuttle facilities (OPF and PCR), and Reference C-8 does the same for the KSC Shuttle cargo facilities (Hangars AE, AM, AO, and S, ESA 60A, DSTF, SAEF 2, VPF, O and C Building and Canister). Both references indicate that airborne particle counting for these facilities will be conducted using automated light scattering techniques. Results obtained for the OPF and PCR by the Shuttle Processing Contractor (SPC, Lockheed Space Operations Company; LSOC) and other facilities by the Shuttle payload integration contractor (McDonnell Douglas Astronautics Company; MDAC) are assumed to be by this method.

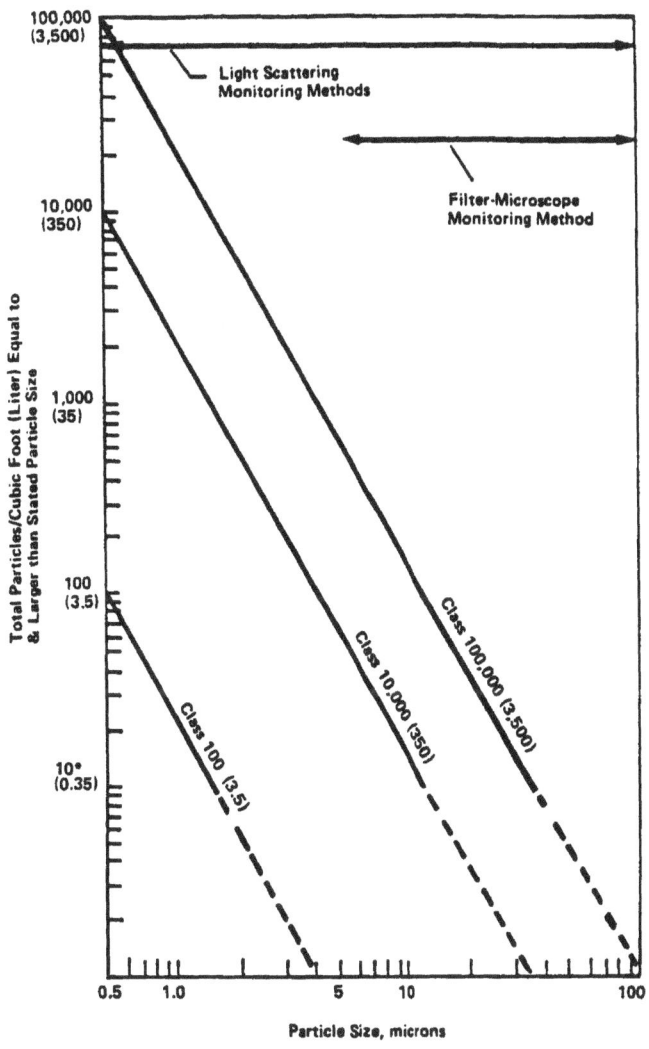

*Counts below 10 (0.35) particle per cubic foot (liter) are unreliable except when large number of samplings is taken.

Figure 2.2-4 Federal Standard 209B Particle Size Distributions (Ref C-4)

For surface deposition, Reference C-3 requests that data be reported in terms of number of particles per ft^2 per 24 hours within given size ranges (50-100 μm, 100-250 μm, 250-500 μm, 500-1000 μm, and >1000 μm). In contrast, Reference C-6 defines surface cleanliness levels in terms of numbers of particles with diameters greater than a given size. These surface cleanliness levels are shown in Figure 2.2-6.

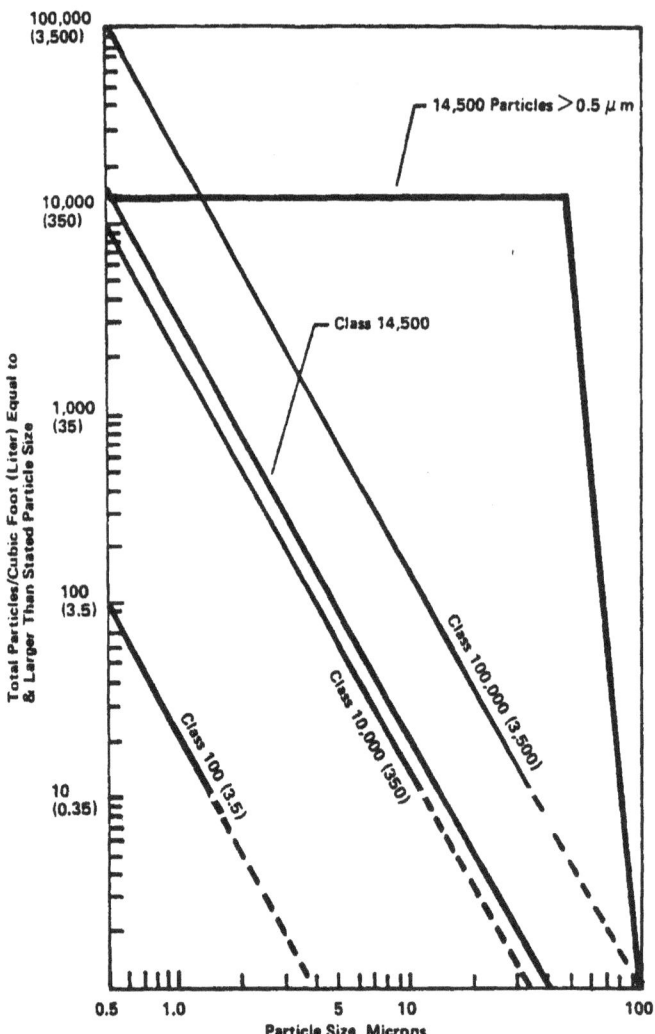

Figure 2.2-5
Effect of Federal Standard 209B Size Distributions

An examination of the data presented herein will show that the surface cleanliness data collected at KSC, and more generally, in most aerospace clean areas, demonstrate different particle size distributions from that shown in Figure 2.2-6. In general, the aerospace data indicates higher relative numbers of large particles in comparison to the curves of Reference C-6. This probably illustrates differences in the facilities, and the activities within the facilities sampled for the two data sets. In general, KSC (and most aerospace) clean areas are not cleanrooms, in that activities such as major component assembly, crane operations and similar industrial type activities are allowed in aerospace facilities.

These all can generate or redistribute particles, including large particles. These types of activities are generally precluded in traditional cleanrooms (Ref. C-4). The origins of the data used to formulate the cleanliness levels of Reference C-6 are not specified. If it was collected in a traditional cleanroom, either away from human activity, or under strict cleanroom discipline, one would expect to see substantially fewer large particles than are typically seen in aerospace facilities. The size distributions of Reference C-6 are similar to those found on vertical or recently cleaned samples, rather than horizontal samples.

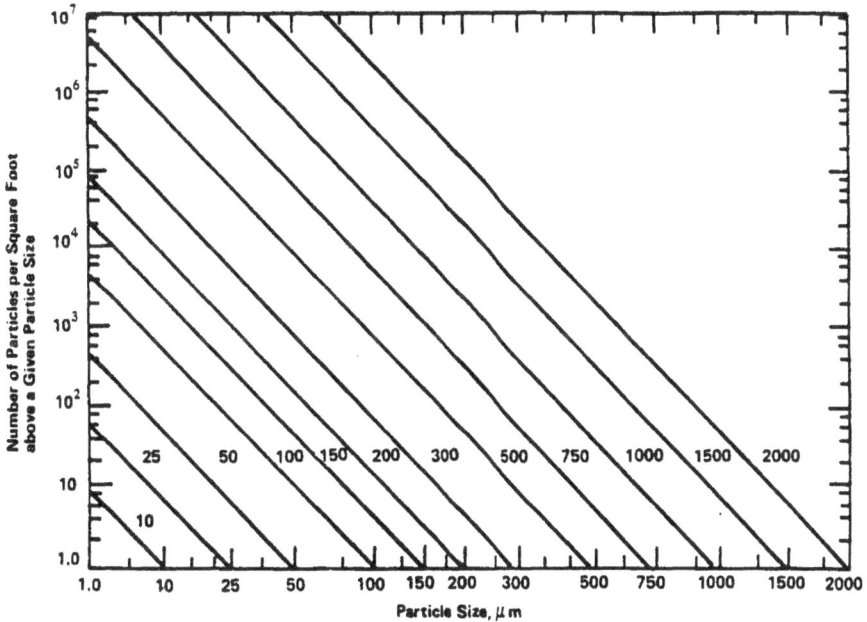

Figure 2.2-6 *MIL-STD-1246A Surface Cleanliness Levels (Ref C-6)*

This difference between the MIL-STD-1246A size distribution and the typical collected data can lead to difficulties in relating the two data sets. Facility and spacecraft specifications often call out surface cleanliness requirements in terms of MIL-STD-1246A levels. Typical KSC fallout data plots traverse several cleanliness levels. Exposure in KSC facilities may result in fairly low general levels of deposition, but with disproportionately high levels of large particles. This can force the apparent cleanliness level to a very high number using the strict definition of MIL-STD-1246A.

Table 2.2-1 listed the environmental requirements for KSC CWAs. For particle fallout, the requirement is listed in terms of maximum MIL-STD-1246A cleanliness level allowable in 24 hours of deposition, with levels of 750 and 1000 being allowed in CWA levels 4 and 5, respectively. A deposition rate corresponding to level 1000 per 24 hours could result in significant degradation over a few days of exposure, if the deposited particles followed the MIL-STD-1246A size distributions. The typical data collected at KSC, however, has a less steep slope than the curves of Figure 2.2-6. This results in a smaller effect from deposited particles for a given surface cleanliness level than if the particles had a MIL-STD-1246A distribution. Section 2.2.4 will describe how size distribution can be included in determining a quantitative measure of surface cleanliness, the obscuration ratio (OR). Section 2.3.2.3 will describe how OR may be used to predict degradation due to particles. In very general terms, based on the data collected to date, daily deposits within the goal of Table 2.2-1 for level 4 facilities should be acceptable for typical integration flows (i.e., ~30 days of exposure) for somewhat sensitive payloads. Highly sensitive payloads would be advised to cover sensitive surfaces whenever possible during exposures. A similar approach should be considered for all payload during long facility exposures.

Reference C-3 suggests three methods of measuring surface particles: 1) flushing the surface with a solvent, filtering the solvent, and sizing and counting the filtered particles; 2) collection of particles on fallout filters with imprinted grids and subsequent sizing and counting of the deposit; and 3) visual inspection of a surface for freedom from visible particles (visibly clean). Levels of visible cleanliness are defined and discussed in Reference C-9. Visible cleanliness criteria are used for surfaces such as the Orbiter PLB and payload surfaces that do not lend themselves to quantitative techniques.

The solvent flush method is sometimes also used on payload surfaces (Ref. C-10). When quantitative data is reported, the method used is often not stated. Evidence that the two techniques give equivalent results has not been presented, although this is the assumption implied in Reference C-3. Since the equivalency of the techniques (and the method used in many cases) is unknown, caution must be used in comparing data from different sources. Both References C-7 and C-8 specify that settling filters and microscopic examinations will be used to determine fallout rates in Shuttle and cargo facilities, so use of these techniques can be assumed in the reports of the SPC and payload integration contractor.

A fallout monitoring technique not included in Reference C-3 has also been used in some KSC facilities. Small circular samples have been installed in the Orbiter PLB and exposed in various ground facilities. They then were removed and shipped to MSFC, where particles were counted using an automated optical imaging system (Ref C-11 and C-12).

A cause for caution in the use of fallout data is the length of time the sample is exposed. In a small test sample, preliminary data seemed to indicate that deposition rates decreased with sample exposure times (Ref. C-13). Whether this reflected changes in the test environment, changes in specimen surface properties, data collection anomalies or some other phenomenon has not been determined.

Another cautionary note is necessitated by the definition of the MIL-STD-1246A cleanliness levels of Figure 2.2-6. The cleanliness level curves define the number of particles equal to or larger than given size. Reference C-3, however, requests that data be reported as the number of particles within specified size ranges, and this is generally the form in which data is received from the laboratory. Consequently, data is sometimes plotted in bandwidth form on the curves of Figure 2.2-6. This is seldom obvious from the curves themselves (only when a curve rises to a higher count for a larger size). Usually, the particle size distribution is such that the resulting error is minor, since there are typically orders of magnitude higher numbers of small than large particles. Occasionally, however, the error can be significant.

One other problem with particle fallout data results from the physical sizes of samples used. Fallout data is usually reported in terms of particles per ft^2 per 24 hours. From Figure 2.2-6, however, 1 ft^2 of a relatively clean surface (level 500) would have approximately 500,000 particles greater than 1 µm in diameter. Manually counting this number of particles is impractical, so typically smaller areas of the sample are counted, and the result is normalized to 1 ft^2 (even for 1 cm^2, a level 500 surface has more than 500 particles larger than 1 µm). The same level 500 example, however, has only one 500 µm particle per square foot. If only 1 cm^2 is examined, on average, the largest particle detected will be 150 µm in diameter. Thus, the fact that most samples are analyzed manually means that it is difficult to get statistically valid counts of large particles.

Some comments are necessary about the use of visibly clean (VC) criteria for PLB and payload surfaces. Such criteria are necessary due to the difficulty of obtaining quantitative particle counts in situ on a payload surface. It would also be impractical to attempt to count particles for each square foot of PLB and payload surface. It is, however, practical to visually inspect all surfaces of the payload and PLB. The surface cleanliness criteria for Orbiter payload and PLB are, therefore, the visibly clean levels defined in Reference C-9. These criteria are summarized in Table 2.2-7. Modifications to these criteria, such as a procedure for sensitive payloads of initial PLB cleaning to VC level 2 in the OPF, and subsequent cleanings and inspections under 50 foot candles of illumination from 2 to 4 feet are currently being considered and implemented. They have been included in some documentation, such as Payload Integration Plans, but an updated version of Reference C-9 has not yet been released.

To date, relationships between the VC levels and quantitative cleanliness levels have not been established. Section 2.2.2.1.9 will describe testing of Orbiter PLB materials recently conducted at Martin Marietta to address this.

The following sections present the available particle data for facilities of interest to Spacelab users.

2.2.2.1.1 <u>Operations and Checkout (O and C) Building Particle Data</u>

Contamination data was collected in the O and C Building during SL-1 (STS-9) processing from August 1982 until April 1983, during horizontal payload processing for STS-41D in July of 1984, and during the first half of 1985.

Table 2.2-7 Visibly Clean Levels and Inspection Criteria (Ref C-9)

Three levels of VC & VC+ SPECIAL requirements are available for the Orbiter payload (cargo) bay, payload canister, & payloads. VC level 1 is baseline as referred to in contractual documentation. VC levels 2, 3, & VC+SPECIAL requirements are optional user services at extra cost & added ground operations time. Inspection criteria for the cleanliness levels follow.

VC Level	Incident Light Level[1]	Observation Distance	Remarks
1	$\geq$ 50-ft Candles	5 to 10 ft	STS Program Standard Service (Baseline)
2	100- to 200-ft Candles	6 to 18 in.	STS Program Optional Service
3	100- to 200-ft Candles	6 to 18 in.	STS Program Optional Service--2x to 7x Power Optical Aid Permitted for Inspection
VC+SPECIAL	100- to 200-ft Candles	6 to 18 in.	STS Program Optional Service--Same Visual Inspection as Level 2 or 3, plus Special Metrology Requirements Specified by User Includes VC + UV, NVR, etc.

The above options are applicable for the Orbiter payload (cargo) bay, payload canister, & payloads at launch sites during premating, mating, & postmating operations.

Note:
1. One foot-candle (lumens per square foot) is equivalent to 10.76 lumens per square meter.

For the SL-1 processing, a tray of optical samples on the Passive Sample Array (PSA) of the MSFC Induced Environment Contamination Monitor (IECM) was installed at MSFC in August 1982, and exposed during transport to and while in the O and C Building until the tray was replaced on April 28, 1983. The replacement tray was exposed in the O and C Building, during transit to the OPF, and in the OPF until it was removed on August 19, 1983.

The trays were covered and returned to MSFC, where an automated image analyzer determined the number of particles present with diameters between 1 µm and 100 µm. Particles larger than 100 µm. were not counted. The average results for these two sets of trays were accumulations of 8.1×10^2 particles/cm^2 for the first set, and 2.0×10^3 particles/cm^2 for the second set (Ref. C-14). The much higher counts for a shorter exposure time of the second set may be attributed, in part, to the heightened activity around the trays as integration activity intensified. The average particle-size distribution collected on the two trays is shown in Figure 2.2-7. The diagonal lines on the figure correspond to cleanliness levels of MIL-STD-1246A. The figure illustrates that the average particle accumulation rates are far below the goal for a KSC level 4 facility (see Table 2.2-1) of accumulations to less than level 750 per MIL-STD-1246A in a 24-hour period. Assuming particle accumulation occurred at a constant rate, the trays collected particles at average rates of approximately level 100 per 24-hours and level 400 per 24-hours, respectively. Such an average may not be a good indication of actual 24 hour deposits. Tests in cleanrooms have indicated that deposition varies widely with time (Ref. C-13).

Additional contamination data for the O and C Building was collected during STS-41D processing. Both air cleanliness and fallout data were collected by contractors for NASA KSC within the time period of July 20 to July 27, 1984. Within this time period, the highest recorded airborne particle count for particles larger than 0.5 µm was 12000/ft^3 (class 12000 per Reference C-4). The recorded particle fallout data was the accumulation of particles equivalent to a cleanliness of less than level 1000 of Reference C-6 in a 24 hour period (Ref. C-5). This is a significantly higher rate of fallout than was evident in the earlier PSA data. Several factors may have contributed to this difference, including specimen handling (PSA samples were shipped to MSFC for evaluation), data collection techniques (automated vs. manual), sample surface characteristics (optical specimens vs. particle count filters), specimen location (SL pallet vs. O and C test stand 2), and proximate activities. Also shown in Figure 2.2-7 is the average particle fallout distribution collected in the period January to June 1985 (Ref. C-15). Airborne particle averages for the same period are shown in Table 2.2-8.

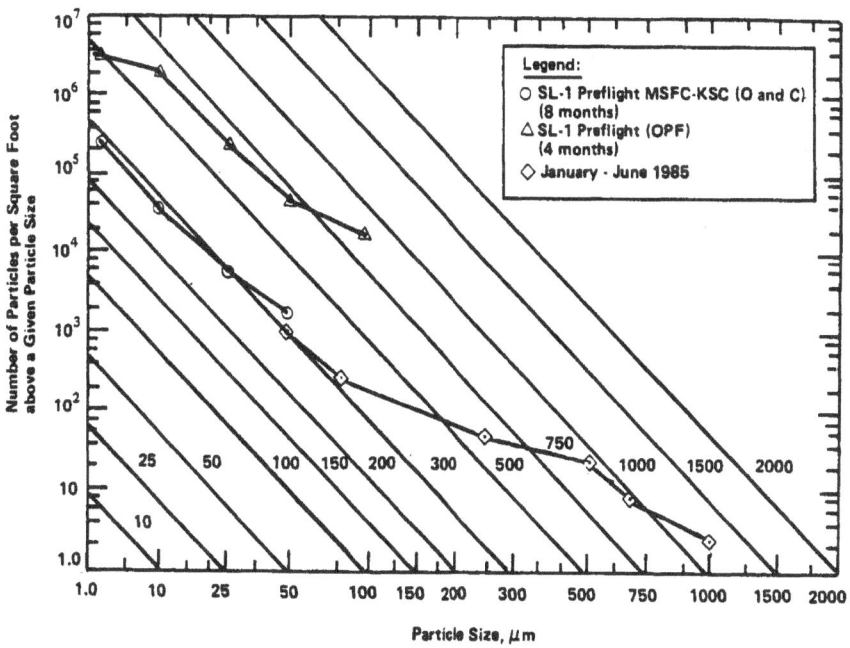

Figure 2.2-7 O and C Building Particle Fallout Data

Table 2.2-8
O and C Building Airborne Particle Measurement

Time Period	Airborne Particle >0.5 μm/ft³
STS-41D	<12,000
Jan to June 1985	35,000 (mean)

2.2.2.1.2 Multi-Mission Support Equipment (MMSE) Canister Particle Data

The MMSE Canister is used to transport vertical payloads from the VPF to the PCR, and horizontal payloads (such as SL) from the O and C Building to the OPF. It is a container with the approximate inner dimensions of the Orbiter PLB. To accommodate both horizontal and vertical payloads, the Canister rotates to these two positions. It is a self-contained vehicle which provides the payload with a CWA level 4 environment of Table 2.2-1.

The first measurements in the Canister were made for the Office of Space and Terrestrial Applications OSTA-1 payload on STS-2. OSTA-1 was transported from the O and C Building to the OPF for installation, and returned from the OPF to the O and C after flight. During the moves, the air cleanliness was generally class 700 to 1500 (number of particles greater than 0.5 μm/ft³), with a peak of class 10,000 (Ref. C-16). Brief excursions above class 100,000 occurred at the same approximate location on both the installation and return trips. The excursions were interpreted as invalid data (Ref. C-17).

For SL-1 (STS-9) processing, one of the ground PSA sample trays was installed in the O and C Building and removed in the OPF. The tray was thus exposed during transport in the Canister. It was also exposed, however, for approximately 4 months in the O and C Building, and briefly in the OPF prior to removal. The results are not indicative, therefore, of the expected fallout for the Canister exposure alone. The results of this exposure, as mentioned previously in Section 2.2.2.1, were an average of 2.0×10^3 particles/cm² for particles with diameters between 1 and 100 μm (Ref. C-14). A comparison of the collected data with MIL-STD-1246A cleanliness levels was shown in Figure 2.2-7. The figure shows that the number of 100 μm particles push the cleanliness level above level 750 (Ref. C-18). Again, however, this level was accumulated over 4 months in the O and C Building as well as during the Canister transport.

Additional Canister data was collected during STS-41D processing of a vertical payload and during STS-51J processing. Table 2.2-9 presents a summary of the environments recorded (Ref. C-5 and C-19). While a typical air cleanliness level of class 5000 was reported for STS-41D, a peak level of class 30,000 was measured during payload transfer operations. Figure 2.2-8 shows particle fallout measurements recorded during rotation of the Canister from horizontal to vertical, and during vertical operation for STS-41B, and all operations for STS-51J (Ref. C-19 and C-20).

2.2.2.1.3 Orbiter Processing Facility Particle Data

The OPF is used for Orbiter turnaround activities between STS flights. As such, major maintenance activities such as Orbiter Maneuvering System (OMS) pod removal and replacement, thermal protection system replacement, hardware cleaning and painting and similar activities will be conducted in the facility. The OPF is also used for horizontal payload installation, which includes SL. While environmental control is provided for cargo activities, the nature of the other activities occurring in the OPF can make the cargo environment somewhat unstable.

Table 2.2-9
Canister Airborne Particle Measurement

Time Period	Airborne Particle $> 0.5\,\mu m/ft^3$
STS-41D	5000 (30,000 Excursion)
STS-51J	< 34,000

As indicated in Table 2.2-3, the OPF as a whole is a CWA level 5 area. Special air conditioning systems provide a CWA level 4 environment to the PLB area while the PLB doors are open. The other activities that occur in the OPF, however, lead to the expectation that level 4 environments will be exceeded in the PLB on occasion.

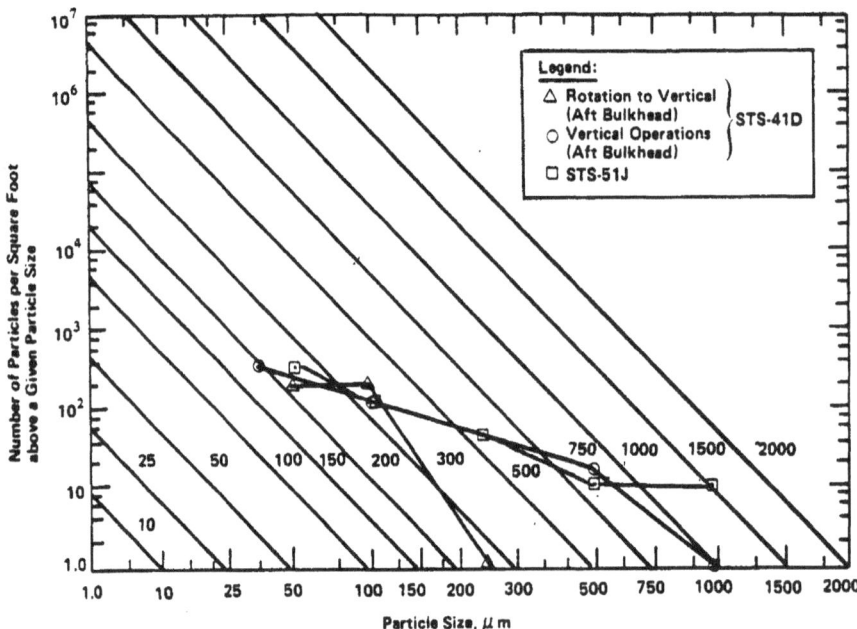

Figure 2.2-8 Canister Particle Fallout Measurements

Measurements of the OPF particle environment were obtained prior to and during the processing of STS-1, STS-2 and STS-3. The measurements indicated that the OPF particle requirements of Table 2.2-1 were all exceeded on some occasions. During STS-3 processing, air cleanliness levels reached class 1,000,000 the day the Orbiter arrived, and were as high as class 320,000 during payload processing. The PLB class 100,000 environment goal was exceeded 22 percent of the time. During STS-3 payload closeout, the 24 hour deposition level in the PLB reached approximately level 2100 (Ref. C-21). Because of these high measurements, an OPF contamination control Program Review Board (PRB) was formed during STS-3 processing. During STS-3 processing, decisions were made by the PRB to upgrade the contamination control provisions of the OPF. The upgrades included addition of a new HEPA filtered airwash for the PLB, replacing the central OPF 80 percent NBS HVAC (National Bureau of Standards Heating, Ventilation and Air Conditioning) filters with HEPA filters, a new vacuum cleaning system for workstands, sealing of the OPF high bay floors, shoe scrubber installation, white painting of structure adjacent to the PLB, installation of permanent airborne particle counters, enclosure of open structures adjacent to the PLB, and paving of selected areas outside the OPF (Ref. C-22).

Since these modifications have significantly improved the OPF environment, detailed data recorded prior to the modifications is not reported herein. If such data is needed, it may be found in References C-12, C-15, C-16, C-21, C-22, C-23, C-24, C-25, C-26, C-27 and C-28.

Subsequent to the OPF facility modifications, a PSA of the IECM was exposed in the OPF for 26 days during STS-9 (SL-1) processing. During this period, the PSA samples collected an average of 2.6×10^3 particles between 1 and 100 μm in diameter per cm^2 ($2.42 \times 10^6/ft^2$). This corresponded to a cleanliness between level 750 and level 1000, and an average 24 hour deposition rate of level 450. As was stated in Section 2.2.2.1.2, such an average deposition rate may not agree with actual 24 hour measurements (Ref. C-13).

OPF particle data has also been collected during all subsequent Shuttle processing flows at the locations shown in Figure 2.2-9 (although data has not always been published). Table 2.2-10 shows the airborne particle counts for STS-41D processing and periods in May and July of 1985 (Ref. C-29). Figure 2.2-10 shows the average fallout distribution for STS-41D and STS-51C processing, and June 1985 (Ref. C-30 and C-31), as well as the total and 24 hour distributions for STS-9 (SL-1). Data for May and July of 1985 were similar to those shown for June.

Table 2.2-10
OPF Airborne Particle Measurement

Time Period	Airborne Particles > 0.5 μm/ft^3	
	High Bay 1	High Bay 2
STS-41D	$<10,000$	
May 24, 1985	<2500	$<16,500$
July 1985	$<68,000$ (Except 450,000 with Doors Open)	$<68,000$ (Except 150,000 with Doors Open)

2.2.2.1.4 KSC Payload Changeout Room (PCR) Particle Data

The PCR is the final room in which the Orbiter PLB may be opened prior to launch. It is an environmentally controlled room (CWA level 4) mounted on the Rotating Service Structure (RSS) connected to the launch mount. The PCR is rotated into contact with the Orbiter and environmentally sealed before the PLB doors are opened. After the PLB doors are reclosed, the RSS rotates the PCR away for launch.

As in the case of the OPF, early measurements in the PCR indicated that contamination could at times rise to unacceptable levels. These concerns reached a head during STS-6 processing when launch delays and extended PCR exposures combined with severe weather, and excessive contamination of the Tracking and Data Relay Satellite (TDRS-A) resulted. Since STS-6, a number of modifications to the PCR have been undertaken. Completed modifications include installation of weather shields, modification of the switch for the hypergolic spill exhaust system, installation of catch plates for the Payload Ground Handling Mechanism (PGHM), an increase in environmental seal pressure and the installation of securing bungee cords, sealing and adjustment of PCR personnel doors, increased lighting in the PLB area, refinishing of the PCR floor, fabrication of payload debris shields, upgrading the PCR wall strength, insulation sealing, automating the PCR environmental monitors, and refinishing ground handling equipment. Other planned changes include increasing the PCR positive

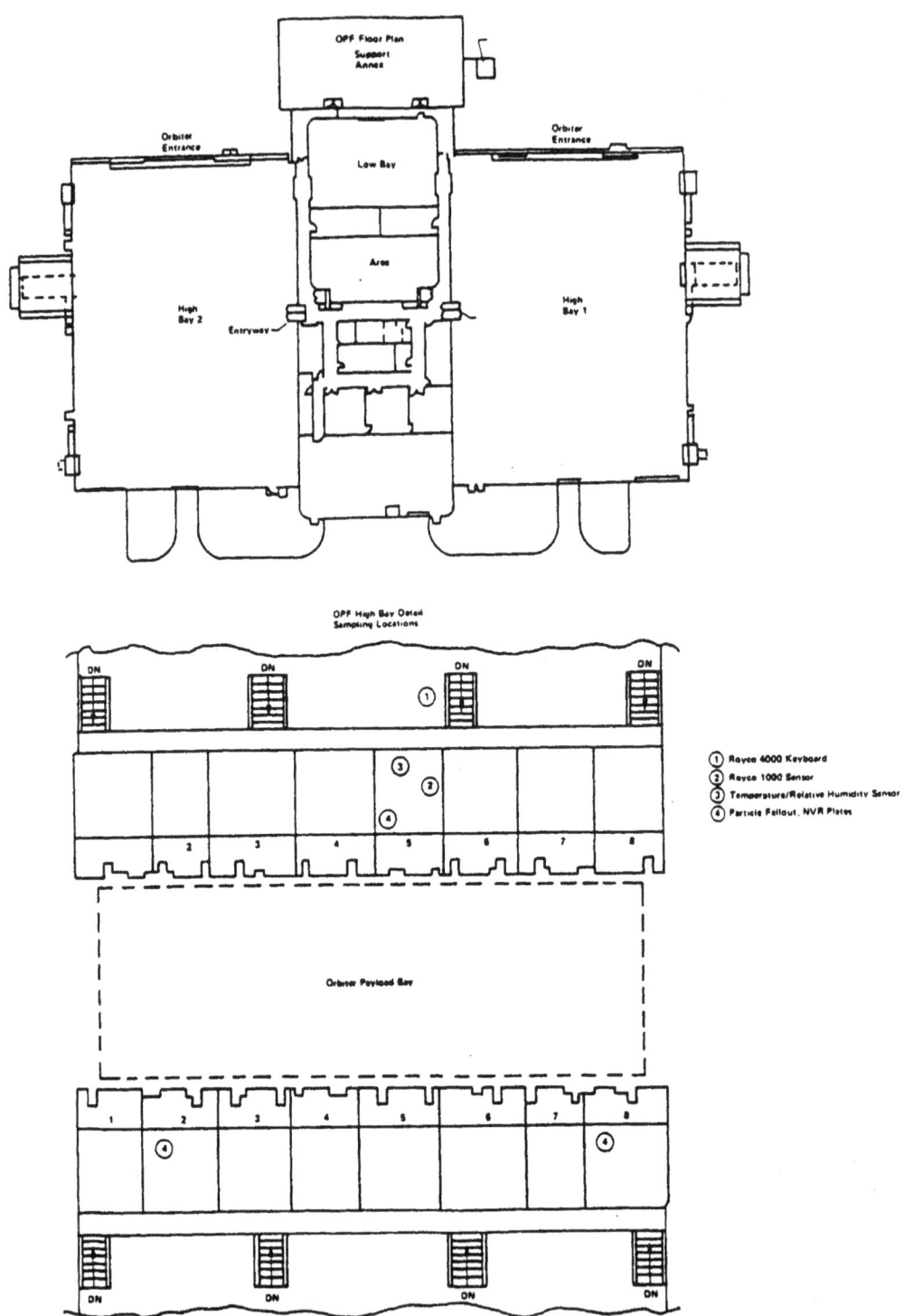

Figure 2.2-9 OPF Floor Plan and Sampling Locations

pressure, enlargement of the PCR anteroom storage area (Ref. C-29), painting of high use areas, addition of smoke and heat detectors, air shower relocation, and the elimination of water intrusion into cable trays (Ref. C-30, C-33 and C-34).

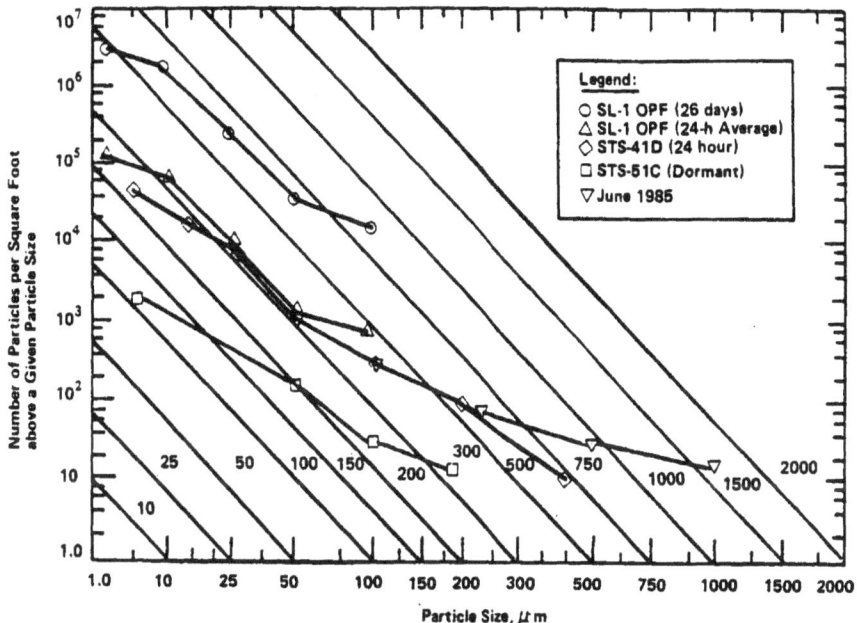

Figure 2.2-10 OPF Average Particle Fallout Measurements

As with the OPF, the PCR modifications have tended to decrease the recorded contamination levels for later STS missions. Data on pathfinder measurements and measurements prior to STS-7 may be found in References C-12, C-15 and C-32.

Airborne particle and particle fallout data has been collected in the PCR during processing of each recent Shuttle flight at the locations shown in Figure 2.2-11. Figure 2.2-12 shows the average 24 hour deposition data recorded for STS-7, STS-51E, and STS-51J (Ref. C-15, C-34 and C-35). The figure shows that the average deposits are greater than level 1000 due to the number of large (600-1000 μm) particles.

PSA samples of the SL-1 (STS-9) IECM were exposed in the PCR. These samples also were in the PLB during OPF to VAB and VAB to PCR transits, rotation of the Orbiter to vertical, launch, orbit, reentry, landing and ferry flight, so it is difficult to reach conclusions about the PCR from these specimens.

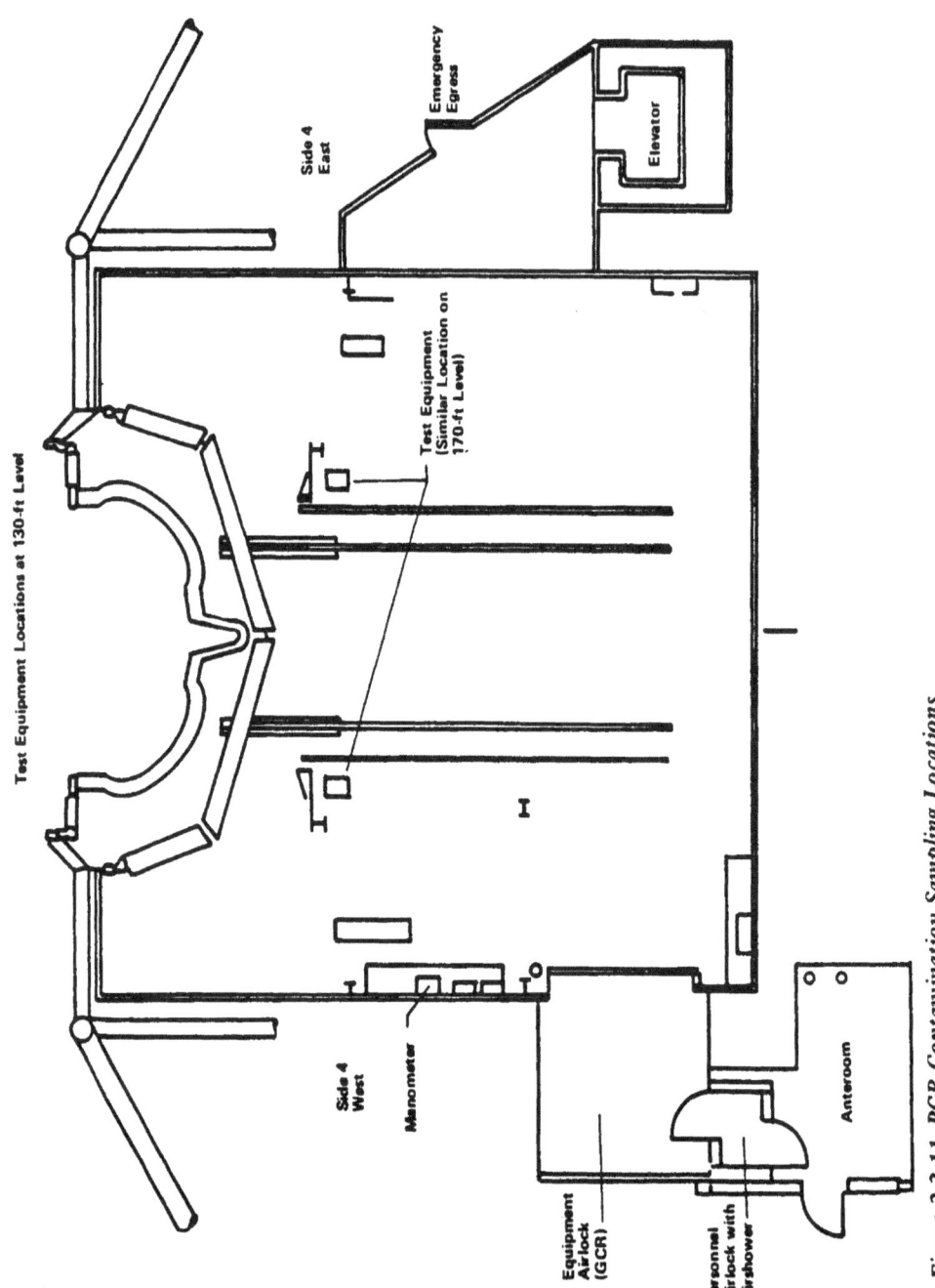

Figure 2.2-11 PCR Contamination Sampling Locations

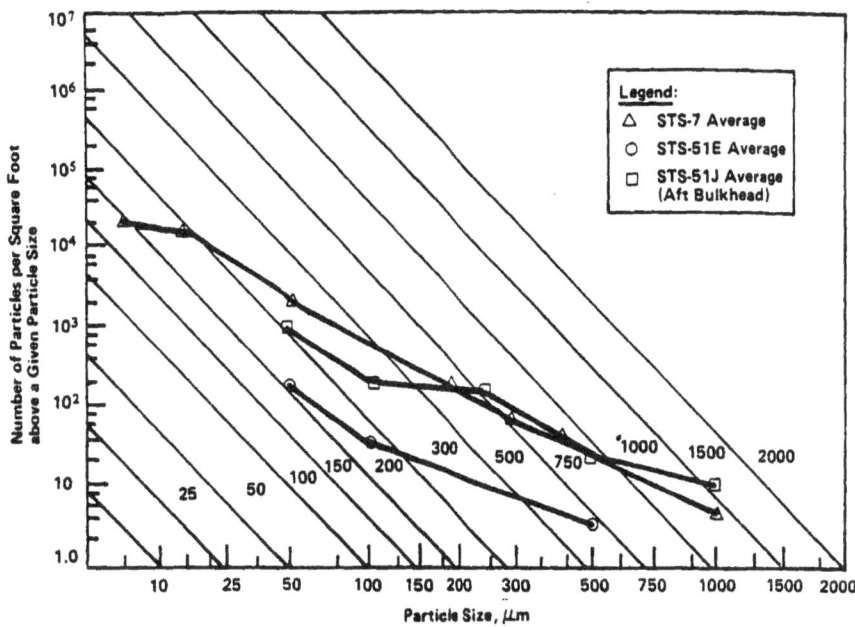

Figure 2.2-12 Average PCR Fallout Rates for Typical STS Flights

Airborne particle counts have been reported for several recent flights. Available data are shown in Table 2.2-11 (Ref. C-15, C-29, C-31, C-36, C-37 and C-38). The data represent the maximum recorded levels. Means were significantly lower.

Table 2.2-11
PCR Airborne Particle Measurement

Time Period	Airborne Particles $>5\ \mu m/ft^3$
STS-7	$<17,000$
STS-51E	$<20,000$
STS-51B	$<7,500$
STS-51F	$<21,000$
STS-51I	$<18,000$
STS-51J	$<10,000$

2.2.2.1.5 STS Ferry Flight Particle Data

STS ferry flight refers to transport of an Orbiter atop the Boeing 747 transport. In general, for the operational STS program, Orbiters will land at the location of their next launch, eliminating the need for ferry flights. Prolonged bad weather or emergency conditions, however, may dictate the use of alternate landing sites. Under these conditions, the Orbiter and payload may land at a site, be safed and mated with the transport and flown to KSC or VLS. Ferry flight must, then, be considered a contingency for any STS mission. It should be noted that ferry flight does not refer to the shipment of integrated SL payloads in a container other than the Orbiter PLB. Such shipment has not yet occurred, and no particle data is available.

Ferry flight particle data was collected by the Passive Optical Sample Assembly (POSA) of STS-1, STS-2, STS-3, STS-4, and STS-9 (SL-1).

The POSA used for the STS-1 ferry flight consisted of 6 sample surfaces. These were a MgF_2/Al mirror, a gold mirror, a 1790Å filter, an ultraviolet (UV) grade fused silica window, a CaF_2 window, and a dielectric (permanent surface charge) Teflon electret. Particle counts were not conducted on the electret. The particle counts for the other 5 samples showed a wide range of both total particle counts and particle size distributions. Figure 2.2-13 shows the average particle size distribution for the five samples (Ref. C-11). No definitive evidence has been presented as to whether the differences in particle counts for the different samples reflect differences in material properties, differences in exposure environment, different relative instrument sensitivities, or some other phenomenon.

POSA samples were installed in the PLB at the landing site (Dryden Flight Research Center for STS-2 and STS-4, WSTF for STS-3), on STS-2, STS-3, and STS-4. The particle size distributions for all ferry flight samples for these three flights were averaged and reported in Reference C-32. The distribution is also shown in Figure 2.2-13. A POSA was also installed at Dryden for the STS-9 (SL-1) ferry flight. Only preliminary data have been published, indicating that only 850 particles/cm^2 were deposited during this ferry (Ref. C-14).

2.2.2.1.6 Orbiter PLB Particle Data

No quantitative measurements of surface deposited particles within the Orbiter PLB are currently available. Martin Marietta has recently conducted experiments relating visibly clean levels of PLB materials to quantitative measures of particle deposition (Ref. C-39). Samples were exposed in cleanroom and aerospace factory environments until a given visibly clean inspection was failed. The samples were then subjected to a particle count, and the particle size distributions were converted to obscuration ratios (the fraction of the surface obscured by particles; see Section 2.2.4.1)

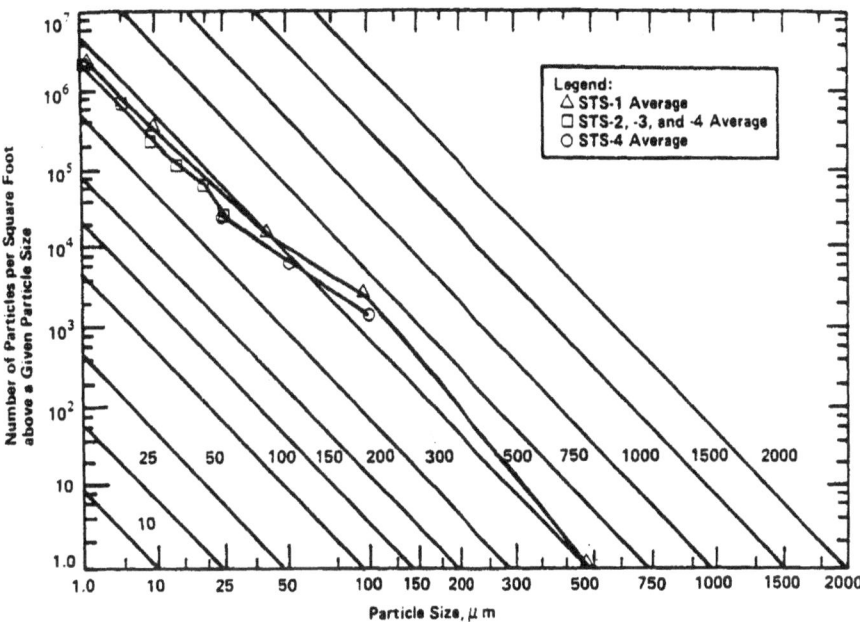

Figure 2.2-13
STS-1, STS-2, STS-3, and STS-4 Ferry Flight Particle Fallout Data

The experiment found that the quantities of particles present on visibly clean surfaces can vary greatly. The "standard" PLB cleanliness level (VC level 1 of Ref. C-9) calls for inspections for cleanliness from 5 to 10 feet under greater than 50 footcandles of illumination. The test found particles to be undetectable at distances of 10 feet. For PLB beta cloth samples failing the VC level inspection at 5 feet, with failure defined as 3 to 7 visible particles on the 0.1 ft^2 samples, obscuration ratios (OR) were found to range from 0.78 to 1.15 percent. Samples with only 1 or 2 particles visible from 5 feet had ORs on the order of 0.25. Inspections from 10 feet could be expected to allow significantly higher deposition levels. Similar tests on the dimpled silver Teflon of the Orbiter radiators shows that it is very difficult to detect particles on the radiators due to their textures. A sample failing an inspection at 4 feet was found to have an OR of 1.04 percent. Additional problems in quantifying radiator cleanliness were encountered due to the presence of static charge on the Teflon, causing particles to clump together around individual dimples.

2.2.2.1.7 Other KSC Payload Facilities Particle Data

Environmental data has also been collected in the VPF, the SPIF and Hangar AE. The VPF is used for the assembly and processing of vertical NASA payloads, the SPIF is an Air Force facility used for processing military payloads, and Hangar AE is an off-line laminar flow cleanroom used for both STS and other payloads. It is included for comparison to non-laminar flow areas. Both the SPIF and Hangar AE are located at the Cape

Canaveral Air Force Station (CCAFS). These facilities are not generally SL processing facilities. Some SL related experiments may fly on non-SL missions, however, and may be processed in facilities other than those of the SL flow. The available data on these other facilities is therefore summarized below.

Particle data for the VPF were collected during pathfinder test flows for the Inertial Upper Stage (IUS) and Payload Assist Module (PAM) upper stages, (Ref. C-23) and for STS-7 (Ref. C-35), STS-8, STS-41B (Ref. C-30), STS-41D (Ref. C-5), STS-41G, STS-51A, and STS-51E (Ref. C-30) payload processing. Figure 2.2-14 shows average 24 hour particle fallout distributions for STS-7, STS-41B and STS-51A. Data For STS-8 was very similar to that of STS-7. STS-41D and STS-51E data generally fell between that for STS-51A and STS-41B. Figure 2.2-15 shows the floor plan of the VPF and the test locations. High airborne particle counts (particles >0.5 µm) were approximately 26,000/ft^3 and 3000/ft for STS-7 and STS-41D processing, respectively. Mean airborne counts were significantly lower. The highest recent level was 19,000/ft^3 during STS-51E processing.

SPIF particle data is available from certification testing, STS-41D and STS-51J processing and the period from October through December 1984. Table 2.2-12 shows the recorded air particle counts for various SPIF locations (Ref. C-5, C-19 and C-30). Figure 2.2-16 shows the recorded particle fallout distributions.

Table 2.2-12 SPIF Airborne Particle Counts

SPIF Location	Airborne Particles $> 0.5 \mu m/ft^3$				
	Stage II Testing	Stage III Processing Flow	STS-41D Payload Processing	Oct to Dec 1984	STS-51J
Integration Cell	500 (South) 1,000 (North)	N/A 6,500	N/A 3,000	<3,000 N/A	<74,000 N/A
Transfer Aisle	100 (South) 5,000 (North)	N/A 8,000	N/A 4,000	<3,000 N/A	<5,000 N/A
Equipment Air Lock	1,000	91,000	11,000	95,000 (Door Open)	N/A
Canister Air Lock	1,500	2,500	N/A	14,000 (Door Open)	N/A
Personnel Air Lock	N/A	N/A	N/A	N/A	N/A
Service Area	100	N/A	N/A	N/A	N/A
Holding Cell	100 (South) 5,000 (North)	N/A 8,000	N/A N/A	N/A	N/A
OAS Room	4,000	N/A	N/A	N/A	N/A

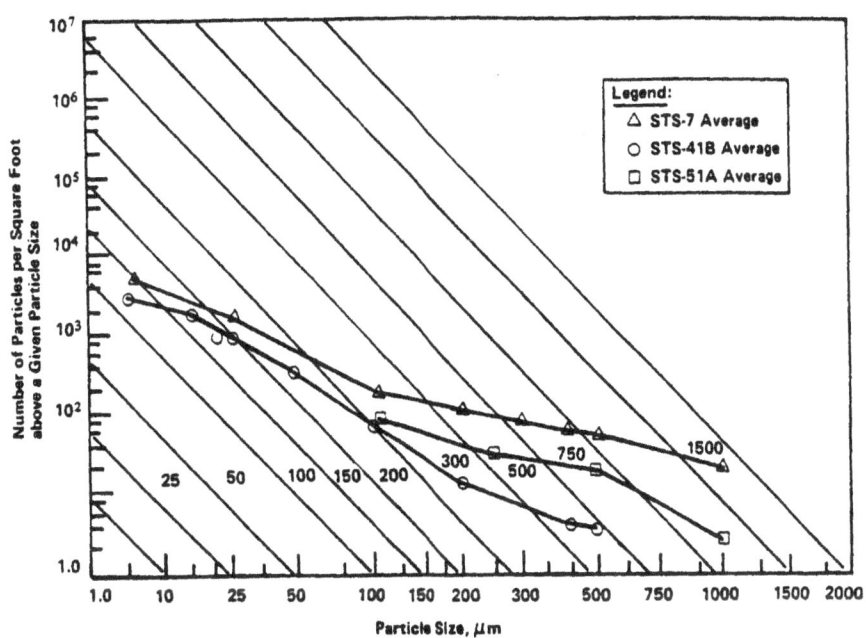

Figure 2.2-14 VPF Average Particle Fallout Distributions

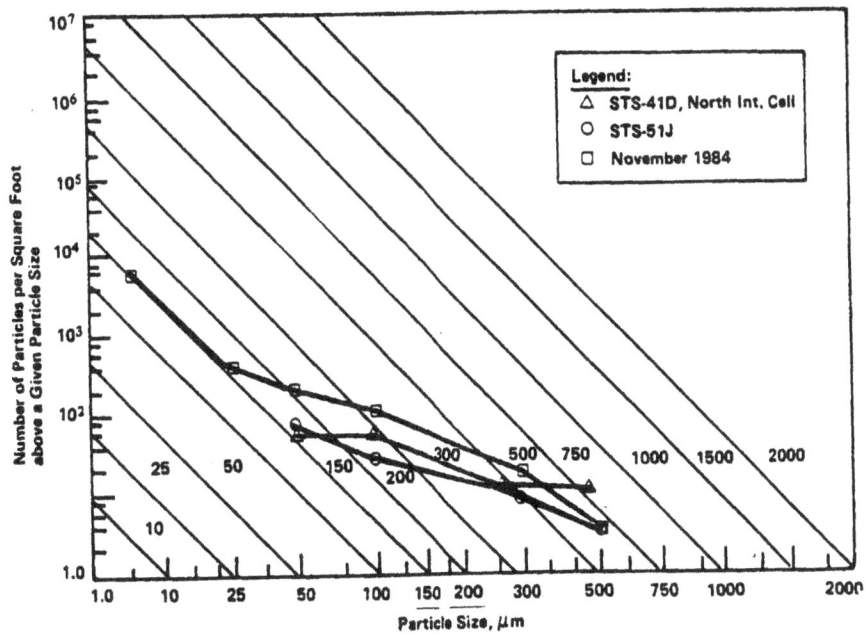

Figure 2.2-16 SPIF Particle Fallout Data

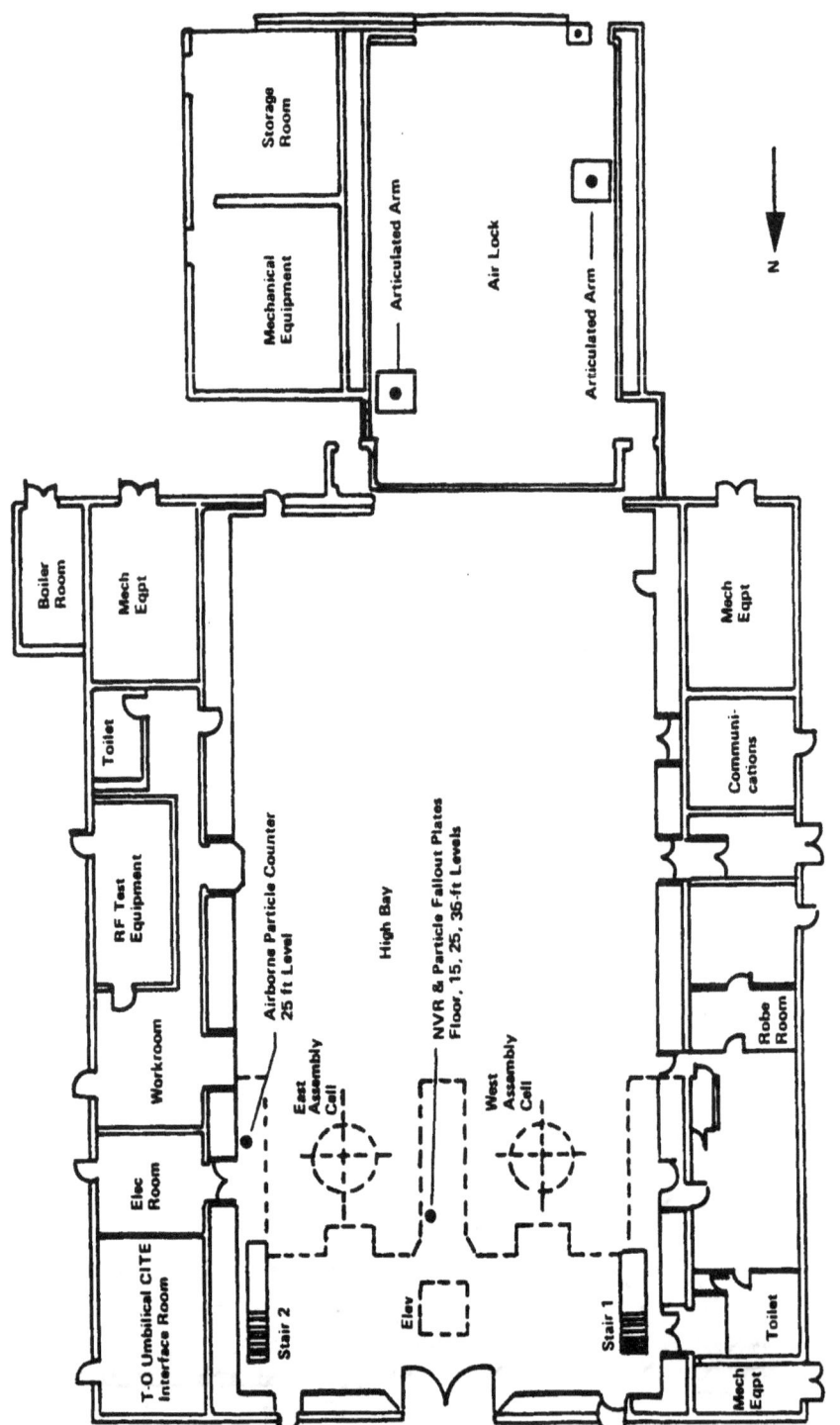

Figure 2.2-15 VPF Particle Test Locations

Data was collected in the Hangar AE laminar flow cleanroom for comparison to the newer non-laminar STS facilities during a study of KSC STS processing facilities (Ref. C-23). Particle fallout data was collected near the HEPA filter bank, near a door to a HEPA filtered non-laminar area, and in a corner away from activity during processing of a non-STS payload. No airborne particle data was collected. The resulting distributions are shown in Figure 2.2-17. The flat portion of the near-filter curve was produced by a single particle larger than 1000 μm. This illustrates how the small sample areas typically counted can lead to overly high sensitivities to large particles in the resulting data.

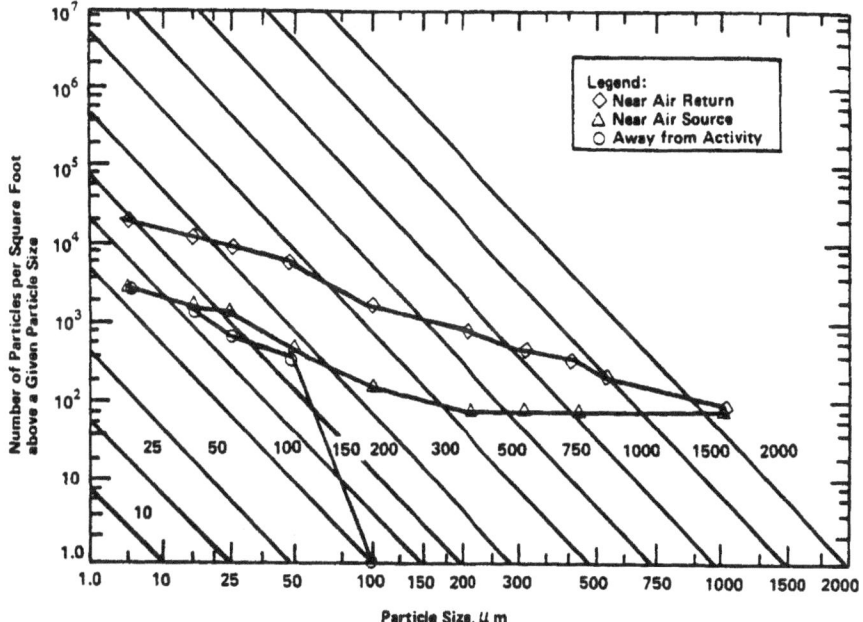

Figure 2.2-17
Hanger AE Cleanroom Particle Fallout Distributions

2.2.2.1.8 VLS Orbiter Maintenance and Checkout Facility (OMCF) Particle Data

The STS facilities at VAFB, including the OMCF, are being completed at this time. Table 2.2-5 presented the environmental requirements for the facilities. Data for verification of these requirements is not yet available. Data will be included in this section as it becomes available.

2.2.2.1.9 VLS Payload Preparation Room (PPR) Particle Data

Some data has been collected in the PPR during facility checkouts. The particle fallout data was generally low, but the applicability to actual payload processing is unknown.

2.2.2.1.10 VLS Payload Changeout Room (PCR) Particle Data

Data for the VLS PCR will also be included as it becomes available.

2.2.2.1.11 Ground Facility Particle Data Summary

Ground particle data has been collected for each of the primary SL ground facilities (O and C Building, OPF and PCR). Data for the OPF and PCR have been collected and reported for a number of STS flights, although data has not been reported for the latest flights. Only limited data has been reported for the O and C Building (2 flights), although as a level 4 facility, continuous monitoring is required. For all three facilities, and the other facilities presented herein, the most recent data is probably most representative of current operational environments, and are generally at such a level so as to be acceptable for typical payloads and integration flows. Highly sensitive payloads and payloads exposed for long periods should be protected (through the use of covers, for example) whenever possible.

2.2.2.2 Ground Facility Molecular Environment

The two most commonly considered ground molecular environmental parameters are the deposition rate of nonvolatile residue (NVR) on surfaces, and the volatile hydrocarbon content of facility air. Two other parameters which can contribute to contamination, and under some circumstances can be considered contamination environments, are facility relative humidity and airborne sodium chloride.

Table 2.2-1 provides requirements for NVR deposition, volatile hydrocarbon content, and relative humidity. No requirement is specified for airborne salt.

To monitor NVR deposition, Reference C-3 specifies that for STS cargo facilities, stainless steel witness plates be exposed to facility atmospheres for a specified period of time, solvent rinsed, and the solvent filtered, evaporated and weighed. The result is to be reported in terms of $mg/0.1m^2/month$. This technique may be assumed unless otherwise specified in all reports by the STS payload integration contractor (MDAC, formerly MDTSCO). Reference C-7 specifies a different NVR monitoring technique for STS processing facilities (OPF and PCR). In these facilities, stainless steel witness plates are also exposed. The plates are then washed with CCl_4 and analyzed by infrared spectrophotometry. The results are reported as N-Hexadecane equivalent. This data is also reported in terms of $mg/0.1m^2/month$. This technique is assumed in reports by the SPC (LSOC).

As was described in Section 2.2.2.1, visual inspections are the standard monitoring techniques for monitoring payload and PLB surface cleanliness. These are primarily particle inspections, since NVR accumulations could reach significant thicknesses before appearing in visual inspections. Work has not yet been done relating VC levels to NVR deposition. The NVR status of PLB surfaces is thus somewhat undefined. Tests were conducted on early STS flights by the Aerospace Corporation in which PLB surfaces were solvent wiped using a 1,1,1 trichloroethane/ethanol solution. The wipes were extracted and the solvent evaporated and weighed.

For volatile hydrocarbons, Reference C-3 requires that a sample of facility air be drawn into an evacuated container. The container is transported to a laboratory where the hydrocarbon content is determined by flame ionization techniques. The result is reported in terms of parts per million (ppm) volume/volume (v/v), methane equivalent. A similar technique is specified in Reference C-7, and may be assumed in reports by both the SPC and payload integrating contractor.

An additional technique for volatile hydrocarbon detection was used for the air sampler of the IECM. Evacuated sample containers were opened in a ground facility. They were then sealed and returned to MSFC, where the samples were analyzed using gas chromatography/mass spectroscopy (GC/MS).

Monitoring requirements for relative humidity in KSC facilities in general are that the device used have continuous recording capability (Ref. C-3, C-7 and C-8). Relative humidity measurements are felt to be well enough understood so as to make the specific monitoring technique non-critical.

References C-3 and C-8, and Table 2.2-1 do not impose airborne salt monitoring requirements on KSC cargo facilities. This parameter is therefore not usually measured in these facilities. Reference C-7 does describe monitoring provisions for the Shuttle processing facilities. Aerosol samples are taken through membrane filters. The filters are placed in deionized water, and the water is analyzed for sodium content by atomic absorption. The results are expressed as grams of $NaCl/m^3$.

The following sections present the available molecular environmental data for STS ground facilities.

2.2.2.2.1 <u>O and C Building Molecular Data</u>

The PSA samples exposed in the O and C Building during STS-9 (SL-1) processing (see Section 2.2.2.1.1) were examined for contamination induced changes in UV transmittance and reflectance. The measurements failed to show any evidence of molecular contamination (Ref. C-14).

Quantitative volatile hydrocarbon, NVR and humidity data were collected by MDTSCO during STS-41D processing in the O and C Building and additional NVR data were collected during the first half of 1985 (Ref. C-15). NVR levels recorded were less than $0.1 \text{ mg}/0.1m^2$/week, or less than $0.5 \text{ mg}/0.1m^2$/month (Ref. C-5). The recorded volatile hydrocarbon level was 2.8 ppm (vol./vol., methane equivalent). The relative humidity ranged between 41 and 45 percent (Ref. C-5). The O and C Building molecular data is summarized in Table 2.2-13.

2.2.2.2.2 <u>MMSE Canister Molecular Data</u>

The MMSE Canister environment was monitored during the STS-2 OSTA-1 payload processing. The recorded volatile hydrocarbon environment remained below 8.4 ppm. Relative humidity was between 30 and 50 percent (Ref. C-17).

Table 2.2-13
O and C Building Molecular Environments

Environment	Time Period	
	STS-41D	Jan to June 1985
NVR	<0.5 mg/0.1m^2/month	0.15 mg/0.1 m^2/month
Volatile Hydrocarbons	2.8 ppm	Not Reported
Relative Humidity	41-45%	Not Reported
NaCl Content	Not Reported	Not Reported

Molecular data were also collected in the Canister during STS-41D, STS-51A, STS-51C and STS-51J processing (Ref. C-5, C-15 and C-30). Canister molecular environments are summarized in Table 2.2-14.

Table 2.2-14
MMSE Canister Molecular Environments

Environment	STS Flight				
	STS-2	STS-41D	STS-51A	STS-51C	STS-51J
NVR	Not Reported	No Measurement	Not Reported	Not Reported	<0.1 mg/0.1m^2/month
Volatile Hydrocarbons	<8.4 ppm	<1.8 ppm	1.5 ppm	44.7 ppm (Painting)	1.3 ppm
Relative Humidity	30 - 50%	36 - 49%	29 - 35%	26 - 35%	40 - 49%
NaCl Content	Not Reported	Not Reported	Not Reported	Not Reported	Not Reported

2.2.2.2.3 <u>OPF Molecular Contamination</u>

Early OPF molecular environmental data was collected during STS-2 processing (Ref. C-17 and C-24). As with the particle data, modifications to the OPF make more recent data more representative of the current situation. OPF volatile hydrocarbon data was also collected on STS-2 by the gas sampler of the MSFC IECM (Ref. C-25) and in the same manner for STS-3 and STS-4 (Ref. C-26 and C-32). The SPC also tested the OPF molecular environment during STS-14 processing, prior to combination of that mission into STS-41D (Ref. C-27). Additional data was collected during June of 1985 (Ref. C-31).

The OPF molecular environment is summarized in Table 2.2-15.

Table 2.2-15 OPF Molecular Environments

Environment	Time Period		
	STS-2	STS-41D	June 1985
NVR	Not Reported	0.31 mg/0.1 m^2/mo	0.3 to 0.7 mg/0.1 m^2/month
Volatile Hydrocarbons	6-9 ppm (15-ppm Painting)	4-6 ppm	2 to 9 ppm (27 ppm Waterproofing)
Relative Humidity	42-48% (85% Excursion)	<50%	Not Reported
NaCl Content	Not Reported	Not Reported	Not Reported

2.2.2.2.4 KSC PCR Molecular Environment

The molecular environment of the PCR was monitored during STS-7 by MDTSCO prior to their selection as payload integration contractor. In this test, the NVR method currently used by LSOC was used (Ref. C-35). Gas samples were taken by the MSFC IECM air sampler in the PCR during STS-9 (SL-1) final closeout (Ref. C-14). The molecular environment has also been reported by the SPC for the STS-9 through STS-51E integration flows (Ref. C-10, C-29 and C-33). The molecular data is summarized in Table 2.2-16 and Figure 2.2-18.

Table 2.2-16 KSC PCR Molecular Environments

Environment	STS Flight			
	STS-7	STS-41B	STS-41C	STS-51E
NVR, mg/0.1 m²/month	<2.4	0.4 - 0.7	0.7 (Avg)	0.15-0.28
Volatile Hydrocarbons	2.1 - 3.9 ppm	2.1 - 3.7 ppm	2.2 - 5.4 ppm	2.7 ppm
Relative Humidity	27 - 60% (Exc to 76%)	20 - 33%	21 - 37%	Not Reported
NaCl Content, mg/m³	0.024 - 2.35	0.9 - 8.6	0.1 - 1.5	0.1 - 18 µg/m³ (sic)

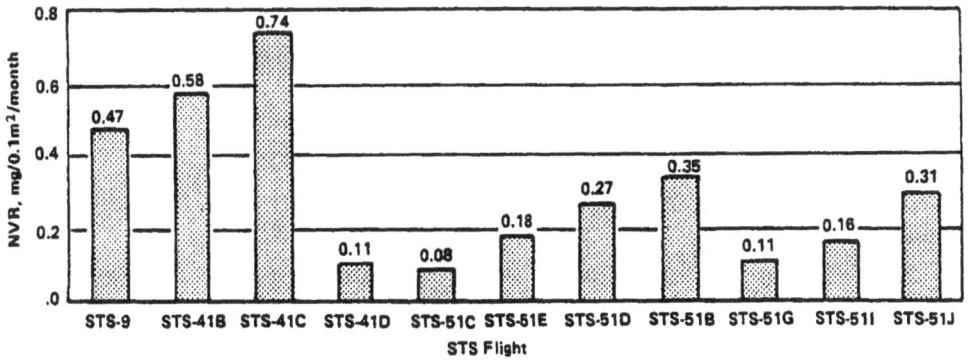

Figure 2.2-18 Average PCR NVR Levels

2.2.2.2.5 STS Ferry Flight Molecular Data

No direct measurements of the molecular environment during post-landing and ferry flight mission phases have been conducted. Indirect measures of changes in optical properties have been conducted using the PSA/ POSA samples described in Section 2.2.2.1.8 and the Optical Effects Module of the IECM (STS-2, -3, -4, and -9, Ref. C-14, C-25, C-26 and C-32). In general, observed changes fall within the sensitivity band of the instru- ment and/or changes are not observed by all samples on a given flight. There is no substantive evidence of significant molecular contamination on STS ferry flights.

2.2.2.2.6 Orbiter PLB Molecular Data

Tests were conducted by the Aerospace Corporation on STS-1, STS-2, and STS-3 where areas of the Orbiter PLB were solvent wiped and the solvent was extracted, evaporated and weighed. The STS-3 measurements were made after the Orbiter had landed and returned to KSC and the OPF. Table 2.2-17 presents the results of these tests (Ref. C-40). The variability of the data probably indicates the presence of localized spots of contamination. Dual readings for the forward bulkhead and right longeron indicate different samples from the same general location. Reference C-40 notes that more standardized cleaning and inspection procedures were implemented subsequent to these measurements (after STS-4).

Table 2.2-17
Orbiter PLB NVR Measurements

Location	STS-1, mg/ft^2	STS-2, mg/ft^2	STS-3, mg/ft^2
Fwd Right Radiator	1.14	0.33	0.15
Mid Right Radiator	0.80	N/A	N/A
Aft Right Radiator	0.34	0.46	N/A
Aft Left Radiator	0.26	0.15	N/A
Fwd Left Radiator	N/A	0.61	N/A
Fwd Bulkhead	N/A	0.48 / 0.80	1.45
Right Longeron	N/A	14.9 / 5.0	1.60 / 0.05
Aft Bulkhead	N/A	0	N/A

2.2.2.2.7 Other KSC Facility Molecular Data

Data on the molecular environment has also been collected in the VPF and the SPIF. The VPF data was collected by MDTSCO during STS-7, STS-41D, STS-41G, STS-51A and STS-51E payload processing. The SPIF data was also collected by MDTSCO during facility certification, STS-41D and STS-51J payload processing (Ref. C-5 and C-20) and October 1984 to January 1985 (Ref. C-30). The NVR measurements for STS-7 processing were conducted using the infrared spectrophotometric technique described in Section 2.2.2.2 (Ref. C-35). All other measurements were as described for MDAC in that section. The molecular environmental data for the VPF is summarized in Table 2.2-18. The data for the SPIF is summarized in Table 2.2-19.

Table 2.2-18 VPF Molecular Environments

Environment	STS Flight				
	STS-7	STS-41D	STS-41G	STS-51A	STS-51E
NVR (mg/0.1m^2/month)	0.06 - 1.8	0.66	0.4	0.3	0.2
Volatile Hydrocarbons	1.6 - 2.1 ppm	2.5 ppm	2.2 - 2.8 ppm	2.0 - 2.2 ppm	2.2 - 2.6 ppm
Relative Humidity	38 - 52%	40 - 46% (Airlock: 43 - 63%)	39 - 43%	Not Available	37 - 52%
NaCl Content	Non Monitored	Not Reported	Not Reported	Not Reported	Not Reported

Table 2.2-19 SPIF Molecular Environments

Environment	Time Period		
	STS-41D	STS-51J	Oct 1984 - Jan 1985
NVR (mg/0.1 m²/month)	<0.2	0.1	0.0 - 1.25
Volatile Hydrocarbons	2 ppm	2 - 5 ppm	2 - 10 ppm
Relative Humidity	37 - 46%	Not Reported	Not Reported
NaCl Content	Not Reported	Not Reported	Not Reported

2.2.2.2.8 VLS OMCF Molecular Data

The environmental requirements of the VLS STS facilities, including the OMCF were presented in Table 2.2-5. Verification of the requirements has not yet been performed. Data will be included in this Handbook as they become available.

2.2.2.2.9 VLS PPR Molecular Data

PPR molecular data will be included in this Handbook as they become available. Some data has been collected during facility checkout, but it was collected while activities not typical of actual processing were occurring.

2.2.2.2.10 VLS PCR Molecular Data

Data for the VLS PCR Molecular environment will be included herein as they become available.

2.2.2.2.11 Ground Facility Molecular Data Summary

In general, the NVR deposition rate is the most important molecular data for payload processors to know. The allowable levels of 1 and 2 mg/0.1m²/month for level 4 and 5 CWAs, respectively, should be acceptable for typical payloads and ground flows. Most of the reported data has actually been well within these allowables. Payloads highly sensitive to very thin layers of molecular deposits should be protected from this environment whenever possible. Protection could include covers and purge systems.

2.2.3 Additional Ground Contamination Data

This section describes additional data not yet published (and, in some cases, raw data not planned for publication), additional data not recorded that is needed in order to more completely define the Shuttle/Spacelab ground contamination environment, and additional data for which collection has been planned.

2.2.3.1 Unpublished Ground Contamination Data

For the purposes of this Handbook, published data is defined to include viewgraph data presented at meetings in addition to the formally published citations. While this definition sometimes includes data presented as preliminary or partial results, and data published only in meeting minutes with limited distribution, it allows the inclusion of data for which no more formal publication will occur. Unpublished data is therefore only that data which has been recorded but not released. It is requested that any reviewer or user of this Handbook who is aware of data not included in the text or bibliography contact the Martin Marietta and/or Marshall Space Flight Center personnel listed in the Foreword of this document.

Considerable quantities of ground contamination data are collected during a typical STS payload processing flow. As shown in Table 2.2-1, Reference C-3 requires that airborne particles, relative humidity, particle fallout and NVR deposition be monitored continuously in KSC level 4 CWAs. Reference C-8, however, requires that environmental data only be reported in the event that requirements are exceeded. The reports are made to the KSC or CCAFS facility manager. In general, facility environmental data has been promptly summarized at frequent Air Force and other contamination working group meetings, and these summaries have contributed significant portions of the data included herein.

In addition to the monitoring by KSC contractors cited above, some monitoring of KSC facilities has also been conducted by other agencies. The Jet Propulsion Laboratory (JPL) has collected ground particle and molecular data on flights STS-6, STS-7, STS-8, STS-41B, STS-41D, STS-41G and STS-51F which were not available for this Handbook edition.

One other set of unpublished ground particle data was collected during STS 51-J. (Ref. C-41) Polyethelene tape as specified in a proposed ASTM procedure for a tape lift test for particulate contamination was provided to LSOC by Martin Marietta. Tape lift samples were collected from various PLB surfaces approximately 3 hours prior to closure of the PLB doors. The results of these tests have not yet been released.

2.2.3.2 Needed Ground Facility Data

The most obviously needed ground facility data is that for the VLS facilities. In the case of the early KSC Shuttle launches, four developmental flights were flown, during which flaws in the contamination control provisions could be identified and rectified. Despite this shakedown phase, significant contamination problems occurred on STS-6, after which further modifications were made to alleviate contamination problems. This is not the approach to be taken at VLS, where the first launch will be an operational military mission. It is important that the lessons learned in bringing the KSC facilities to operational status be heeded, and that appropriate environmental data be collected prior to facility first use.

Other needed data includes: 1) the NVR deposition level in the PLB for current operations (the amount of PLB NVR has an impact on the flight molecular environment, and may be representative of payload surfaces at launch); and 2) a comparison of the various techniques currently used for NVR monitoring at KSC.

An entire set of data similar to that collected on STS-51J and described above is needed to accurately quantify the Orbiter PLB cleanliness at launch. There are currently no reliable data on actual PLB cleanliness, nor is there a method for quantifying the cleanliness of a surface that passes a VC-1 inspection from 10 feet.

2.2.3.3 Planned Ground Facility Data

Under the requirements of Reference C-3, the monitoring of KSC facilities during Orbiter and payload processing operations will continue. In addition to these, some other ground facility monitors are planned. NASA has recently incorporated a set of standard witness plates on the STS aft bulkhead cover used during ground processing which is removed prior to launch.

Several planned tests described in the first edition of the Handbook are no longer planned. The fifth flight of the IECM was scheduled for the STS-51F/SL-2 flight. Weight restrictions forced the removal of the IECM from the manifest. No further flights are anticipated. Tests at JSC similar to the Martin Marietta Visibly Clean quantification tests did not yield useful results. Some testing may continue at the NASA WSTF. Finally, a comparison of the various NVR test methods currently in use was planned by Martin Marietta. The technique used by one of the KSC contractors has not been obtainable, since the technique is considered proprietary, so this comparison has not been possible. An effort is currently underway by the American Society for Testing and Materials to standardize NVR tests, and this may lead to a more uniform data base in the future.

2.2.4 Ground Facility Analytical Tools and Models

Section 2.2.2 presented the currently available facility contamination data for STS-related facilities. This section provides descriptions of models and examples for translating the ground data into predictions of contamination of Spacelab experiment surfaces. Current models will predict levels of surface deposition from ground facility sources. A discussion of the prediction of the effects of the comtaminant deposits will be deferred to Section 2.3, Flight Contamination Environment since the launch environment must be considered before final on orbit deposition levels may be determined.

2.2.4.1 Ground Particle Analytical Tools

The information available on STS facility particle environments generally includes the minimum environments the facility is required to provide, and at least some data on the actual environments provided. The amount of data available for given facilities varies greatly. Data may be available for a facility from only one or a few STS missions, or from most flights. Data may be summaries of peak levels, or compilations from which averages may be inferred. Data on facility airborne particles are reported in various forms such as number of particles per ft^3 greater than 0.5 µm, and additionally (in some cases) numbers greater than 5 µm. Surface deposition rates may be in terms of total particles collected larger than a given size, numbers of particles in a given range, or graphical representations of complete particle size distributions. In contrast, the most interesting information for a user of the STS as a launch vehicle is the type of contamination present on the surfaces (and, in some cases, in the field-of-view) of his equipment when the equipment must perform on orbit. The quantities and types of data available from appropriate ground facilities determines the types of analysis tools needed to predict particle deposition levels. This section describes how the various types of available data may be translated to total predicted ground deposits. Section 2.3.2.3 will describe how total deposition on orbit may be inferred from ground deposits, and how predictions of Orbiter PLB particle generation behavior may be made. After the orbital levels are determined, the user can utilize the effects information of Section 2.3.2.1.4 to determine if his equipment is compatible with the predicted environment, or whether some of the prevention techniques of Sections 2.2.5 and 2.3.2.4 are necessary.

2.2.4.1.1 <u>Obscuration Ratio from Deposited Particle Size Distributions</u>

The most detailed particle deposition data generally available from STS facilities is a distribution of particle sizes and numbers deposited on a surface in a 24-hour period. A more useful parameter for evaluating surface particle contamination is the fraction of total surface area obscured by the deposited particles, termed the obscuration ratio. A technique has therefore been developed to convert a particle size distribution to an obscuration ratio. Section 2.3.2.3 will describe how the obscuration ratios of the various Orbiter PLB surfaces at launch may be used to predict an obscuration ratio of a given surface on orbit, and how obscuration of various surfaces affect their performance.

The important parameter in determining obscuration ratio for a surface is the number of particles per unit area present as a function of particle cross-sectional area. The particle size used in developing particle size distributions is typically the greatest linear dimension of a given particle. Thus, a spherical particle 500 µm in diameter and a fiber 500 µm in length appear on the final distribution as equivalent, even though they may have significantly different cross-sections. To assess how cross-section varies with particle size, a microscopic evaluation of the ratio of particle width to length (w/d) was conducted on particles collected in operational cleanrooms at Martin Marietta (Ref. C-42). The evaluation indicated that for particles with largest dimensions less than approximately 70 µm, the width to length ratio is usually on the order of 0.65, with 1.0 being spherical. For particles with dimensions greater than 70 µm, the particles become increasingly fibrous with particle size. The data collected is shown in Figure 2.2-19. Areas for particles are calculated assuming that particles are cylindrical with hemispherical ends. Particles with width to length ratios of 1 have circular areas, while those with small width to length ratios are nearly rectangular.

Once a particle size distribution and a function defining the variation in cross-section with size is available, the area obscured by the presence of particles may be obtained directly by integrating particle cross-section over the particle size range. This integration has been performed at Martin Marietta for several linear (on log normal plots) particle size distributions of various slopes. A mathematical function was then developed which approximates the integral results for particle size distributions with slopes ranging from -0.333 to -0.5 (typical of measured KSC facility distributions). Thus, for a linear particle size distribution, and considering the variation of particle cross-section with size, obscuration ratio may be determined from:

$$OR = N(TNP) \qquad 2.2-1$$

where

OR = obscuration ratio;
TNP = total number of particles/ft^2 larger than 5 μm;
N = 10^A, where A = $-(35.4[Log(-SLP) + 2.294])^{1/2}$; and
SLP = slope of the particle size distribution when plotted on a log-normal scale.

Example 2.2.4.1.1	Calculation of daily surface obscuration based on STS-41B worst average 24-hour fallout for the PCR.
Solution:	Worst daily average STS-41B 24-hour particle deposition is shown in Figure 2.2-20. Linear regression gives a line of slope -0.4504 with 5687 particles larger than 5 μm deposited per day for this curve. The curve and the linear approximation are shown in Figure 2.2-20. Applying Equation 2.2-1, $$OR = 2.8 \times 10^{-5}/day.$$

This method calculates obscuration ratios based on particles larger than 5 μm. Reference (C-43) indicates that OR can be strongly influenced by particles smaller than 5 μm. This analytic conclusion is based on particle size distributions parallel to (and some steeper than) those from MIL-STD-1246A. For those types of slopes, small particles can strongly affect obscuration ratios. For slopes more characteristics of actual fallout data, however, particles less than 5 μm have a much smaller effect. Slopes of sub 5 μm, particles tend to naturally be small, since settling rates for these are small (although the resistance to cleaning small particles possess tends to counteract the lower settling rate impact). The error induced in calculated obscuration ratios by neglecting particles between 1 and 5 μm ranges from 1 percent for slopes parallel to typical 5 to 50 μm data, to 8 percent for slopes parallel to MIL-STD-1246A distributions. Actual size distribution data for particles smaller than 5 μm are not typically available, due to the difficulty in counting these sizes. Currently, it is recommended that obscuration ratios calculated base on size distributions for larger than 5 μm particles be used, pending data showing that this approach results in unacceptable inaccuracy.

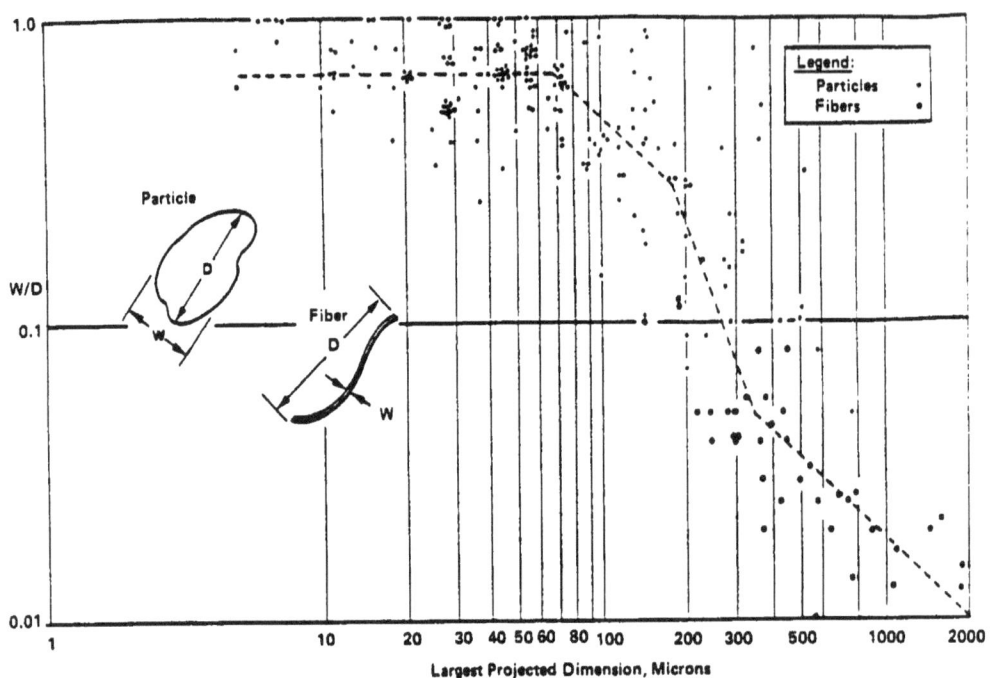

Figure 2.2-19 Particle Shape as a Function of Particle Size

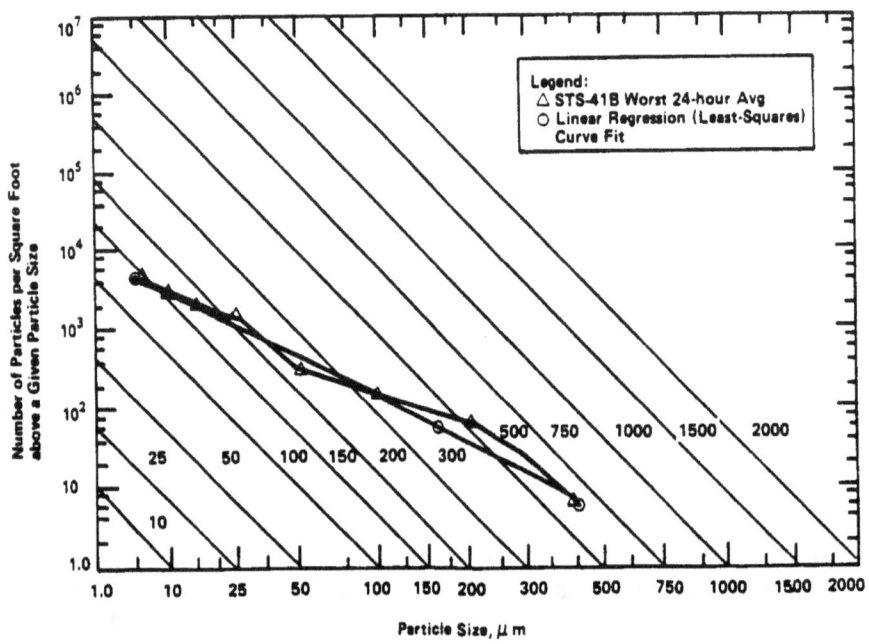

Figure 2.2-20
STS-41B PCR Particle Fallout Data and Linear Data Approximation

The above method requires counts of particles larger than 5 µm. Most recent KSC fallout data, however, does not include information on particles smaller than 50 µm. A second technique has been developed recently by Raab of Martin Marietta. This technique relies on an "average" particle size distribution derived from approximately 60 sets of particle fallout data for the OPF, SPIF, VPF, and PCR. The number of particles larger than 50 µm is then used to determine where this average size distribution falls on a MIL-STD-1246A plot. Equation 2.2-1 is again used to calculate OR, but in this case:

TNP = total number of particles/ft^2 larger than 50 µm; and
N = 6.58×10^{-8}.

It should be noted that this second technique relies on the size distribution being "typical". Unusual distributions will have actual ORs not in agreement with those calculated by this technique.

2.2.4.1.2 Obscuration Ratio from Total Fallout Count

In some instances, particle fallout distributions are not presented, and fallout data is simply presented as the number of particles present larger than 5 µm. The technique used to address this situation provides a general obscuration ratio based on a representative particle distribution curve. The technique will also be used in the next section to predict obscuration rates from air cleanliness data.

Reference C-44 reviewed particle size distributions for five previous studies of cleanroom particle fallout, as well as those of MIL-STD-1246A (Ref. C-6). When the measured data were normalized to show relative populations and fit to straight lines, the slopes of the lines fell between -0.311 and -0.557 when plotted on the log normal scale of Reference C-6. Four of the five data sets fell between slopes of -0.311 and -0.380. The slope of Reference C-6, on the other hand, is -0.926.

If the one slope not between -0.311 and -0.380 is discarded, the average for the four remaining slopes is -0.34. Since a smaller absolute slope value indicates a greater relative number of large particles, obscuration ratio for a given number of particles increases as the slope approaches zero. A slope of -0.333, while representative of the four data sets used, is somewhat more conservative than the numerical average value. This is the value of particle distribution slope selected by Martin Marietta in developing a generalized obscuration ratio calculation technique (Ref. C-42).

When Equation 2.2-1 is applied using a slope of -0.333, the curve of Figure 2.2-21 results. This figure presents obscuration ratio as a function of total number of particles larger than 5 µm, assuming a linear particle size distribution slope of -0.333 and taking into account the variation in particle cross-section with size. The curve can then be used to predict obscuration or obscuration rates when only total particle counts are available.

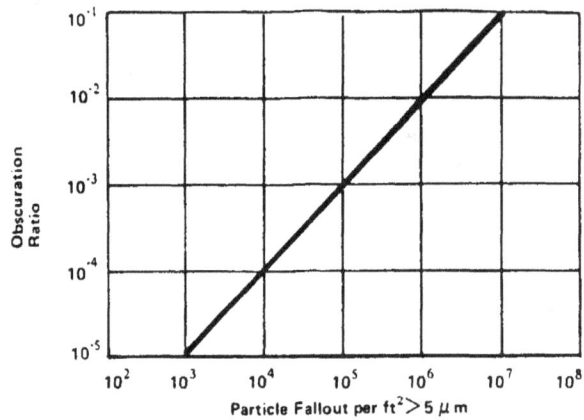

*Figure 2.2-21
Obscuration Ratio versus Number of Particles per ft²
> 5 μm Using Experimental Shape Factor*

Example 2.2.4.1.2:	Calculation of obscuration ratio from total particle fallout counts, using average total PCR counts from Reference C-23, STS-4.
Solution:	Reference C-23 reports that the average 24-hour fallout count for the PCR during STS-4 operations subsequent to the MMSE canister arrival was 7309 particles/ft². For this total particle count, Figure 2.2-21 gives an obscuration ratio of 6.65×10^{-5} per day.

2.2.4.1.3 <u>Particle Fallout Predictions from Air Cleanliness Data</u>

In some cases, representative witness plate fallout data is not available for a facility of interest. If air cleanliness data or requirements are available, the number of particles per ft^3 larger than 5 μm may be used to predict the rate of particle deposition on a surface.

Reference C-45 compiled air cleanliness and particle fallout data from a number of earlier cleanroom studies. The fallout data used was the total number of particles collected in 24 hours per ft^2 larger than 5 μm. The air cleanliness data used was the average number of particles per ft^3 larger than 5 μm. The data used in the referenced study are plotted in Figure 2.2-22. This figure represents the Hamberg particle fallout model, named for the author of the study. The author performed a nonlinear regression analysis using the logarithms of the fallout rates and developed the following expression for fallout as a function of air cleanliness.

$$n = 2.851(10^3)N_c^{0.773} \qquad \qquad 2.2\text{-}2$$

where

n = number of particles >5 μm settled/ft²/24 hours; and
N_c = number of particles >5 μm /ft³ of air.

This equation represents the mean curve of Figure 2.2-22. Hamberg recommends that this and other curves of the figure be used as follows:

a) For conventional type cleanrooms, with 15 to 20 air changes/hour, use the mean fallout rate;

b) For cleanrooms with considerably less than 15 to 20 air changes/hour, approaching still air, use the maximum 95 percent confidence limits to obtain conservatively high fallout rates; and

c) For cleanrooms with directional velocities near 90 feet/minute, approaching laminar flow type rooms, use the minimum 68 percent confidence limits to obtain conservatively high fallout rates.

KSC payload facilities generally fall in the 15 to 20 air changes per hour category. In a study of KSC facility environments and fallouts during the first four STS ground flows, the correlation was generally found to fall near the mean curve of Figure 2.2-22, as is suggested in the above groundrules. Thus Equation 2.2-2 may be used directly, or the mean curve from the figure may be used to predict particle fallout rates from air cleanliness data. The technique described in Section 2.2.4.1.2 for Figure 2.2-21 can then be used to convert the predicted fallout count to an obscuration ratio.

Example 2.2.4.1.3:	Based on the peak airborne particle count of STS-41B processing in the PCR, calculate the daily fallout rate and obscuration rate. (See Section 2.2.2.1.6)
Solution:	Reference C-10 reports that during PCR main door opening and canister arrival activity, airborne particle counts of 14,500/ft³ for particles larger than 0.5 μm were recorded. Assuming an airborne particle distribution from Fed. Std. No. 209B, 102 particles/ft³ larger than 5 μm would be predicted from this count. From Figure 2.2-22, the mean curve gives a deposited particle prediction of 95,000 particles/ft² for this count. Figure 2.2-21 translates this prediction to 0.1 percent obscuration per day.

2.2.4.1.4 <u>Prediction of Obscuration Ratio from Visibly Clean Inspections</u>

Section 2.2.2.1.9 described the results of tests to quantify NASA visibly clean levels. The tests provide quantitative correlations for inspections of surfaces from up to five feet, but indicate that the ten foot maximum inspection distance for the Shuttle Standard (VC-1) cleanliness level may be ineffective for particle detection. For inspections from five feet, a typical OR of 1 percent is indicated for Beta Cloth (PLB liner material) surfaces, and 2 percent for dimpled silver Teflon (Orbiter radiators). Mirrored surface ratios as low as 0.25 percent were also seen in the test. Data is unavailable to quantify surface cleanliness after inspections from ten feet, but obscuration ratios 4 times higher than those measured in the test may be reasonable. Values of 2 percent obscuration for VC-1 Beta Cloth surfaces and 4 percent for dimpled silver Teflon are suggested for total particle population analysis (Standard Orbiter Cleanliness)

2.2.4.1.5 Cleanliness and Obscuration on Vertical Surfaces

In testing at AEDC (Ref. C-46), the ratio of vertical to horizontal deposition of particles was found to average 0.2 percent. In similar testing at TRW (Ref. C-47), an average as high as 10 percent was reported. No explanation for this wide disparity has been offered. In two recent tests at Martin Marietta, average ratios of 5 percent (Ref. C-13) and 3.2 percent (Ref. C-38) were recorded. Obscuration ratio was also measured in the second test, with vertical surface obscuration 1.2 percent of that found on horizontal surfaces. This is the ratio currently used in contamination analyses at Martin Marietta. Given the ratio of vertical to horizontal deposition, vertical predictions may be made directly from the results already described in Sections 2.2.4.1.1, 2.2.4.1.2, and 2.2.4.1.3.

2.2.4.1.6 Integrated Ground Particle Analysis

As will be described in Section 2.3.2.3, the critical parameters for predicting the particle contamination levels of STS payloads on orbit are the payload critical surface location and orientation, and the obscuration ratios present at launch for each of the Orbiter PLB and payload surfaces. In general, the obscuration ratio of a surface at launch is given by:

$$OR = OR_0 + \dot{OR}(t)t \qquad 2.2-3$$

where

OR_0 = obscuration ratio of surface after last cleaning prior to launch;
$\dot{OR}(t)$ = time (or facility) varying obscuration ratio change rate; and
t = time from last cleaning to launch.

Section 2.3.2.3 will describe how particles may be dislodged and will migrate during launch. Thus, the cleanliness of a particular instrument at launch may become secondary in comparison to the cleanliness of the huge vertical surface areas of the Orbiter PLB (liner, doors, and radiators). The general application of Equation 2.2-3 is therefore to calculate the obscuration ratios of the large surface areas of the Orbiter. Table 2.2-20 summarizes the large Orbiter PLB surface areas (Ref. C-48).

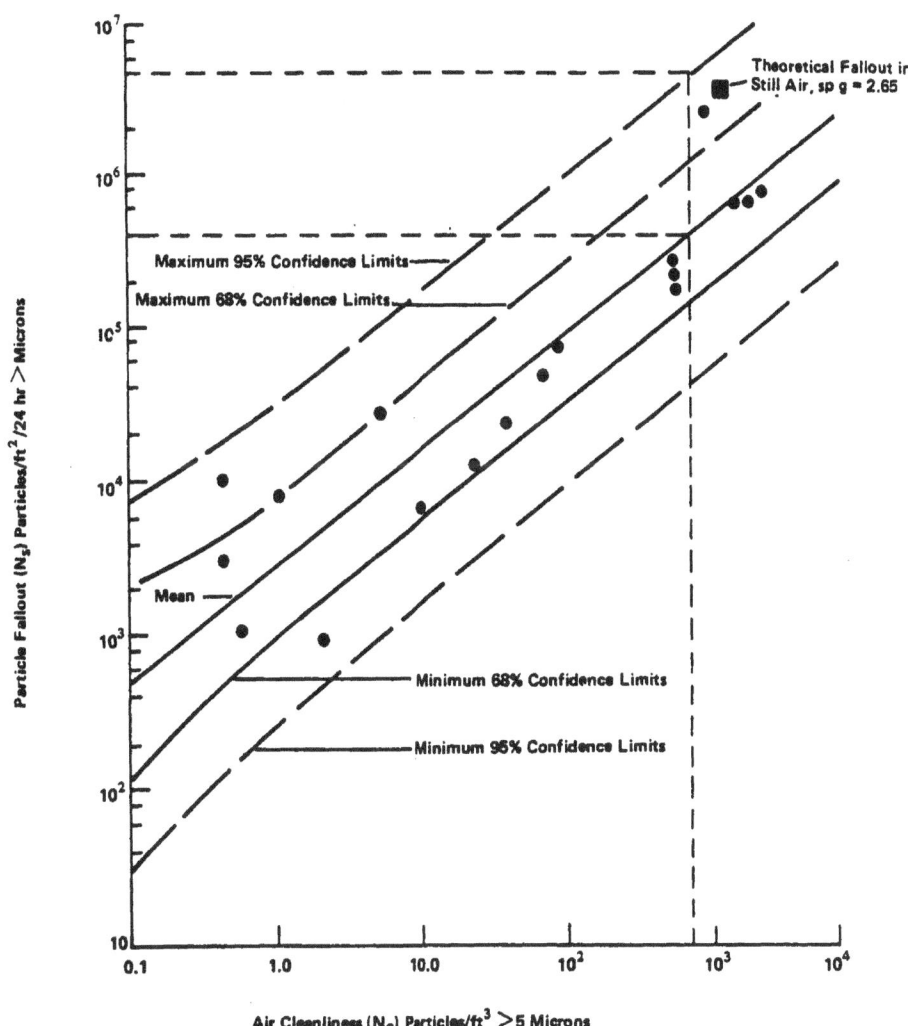

*Figure 2.2-22
Hamberg Model of Particle Fallout Rates with Originating Data
Points (Ref C-34)*

Table 2.2-20 Orbiter PLB/SL Surface Areas

Surface	Total Area (cm^2)	Orbiter Horizontal		Orbiter Vertical	
		Horizontal Area (cm^2)	Vertical Area (cm^2)	Horizontal Area (cm^2)	Vertical Area (cm^2)
PLB Liner	1.31x10^6	6.55x10^5	6.55x10^5	0	1.31x10^6
Doors/Radiators	1.31x10^6	1.31x10^6	0	0	1.31x10^6
Bulkheads	3.28x10^5	0	3.28x10^5	1.64x10^5 (Same Inverted)	0
SL Long Module	1.25x10^6	2.41x10^5 (Same Inverted)	1.02x10^6	1.43x10^5 (Same Inverted)	9.65x10^5
SL Short Module	7.84x10^5	1.01x10^5 (Same Inverted)	5.43x10^5	1.43x10^5 (Same Inverted)	4.98x10^5
SL Pallet	1.16x10^5	5.8x10^4	5.8x10^4	0	1.6x10^5

2.2.4.2 Ground Molecular Analytical Tools

Data for up to four ground molecular environments were presented in Section 2.2.2.2: NVR deposition, airborne volatile hydrocarbons, airborne NaCl, and relative humidity. Some portion of ground NVR deposited on PLB SL surfaces will remain as a contaminant on surfaces in the STS orbital environment, while the remainder of the ground NVR will be outgassed as a potential contaminant for other surfaces or fields-of-view. Ground volatile hydrocarbon content may influence NVR deposition. Airborne NaCl can influence contamination generation from corrosion and may deposit as a permanent contaminant itself. NaCl deposition is included as NVR in gravimetric NVR analyses, but not in infrared spectrophotometry analyses looking for hydrocarbons. Relative humidity also influences corrosion rates, and airborne water may be absorbed by hygroscopic materials and be re-emitted on orbit as a field-of-view contaminant.

While the above data are routinely collected, models do not currently exist for utilizing the airborne data. Ground molecular analysis techniques are limited to prediction of surface NVR levels at launch.

The analysis technique for predicting NVR levels is similar in form to that used for obscuration ratio:

$$NVR = NVR_0 + \dot{NVR}(t)t \qquad 2.2\text{-}4$$

where

NVR = deposited nonvolatile residue at launch;
NVR_0 = nonvolatile residue remaining after last cleaning operation;
$\dot{NVR}(t)$ = time (or facility) dependent nonvolatile residue deposition rate; and
t = time from last cleaning to launch.

The application of this method is considerably different from that for particulate obscuration ratio, however. With obscuration, the total particle population is important because particles are readily excited and migrate during launch. With NVR, the standard analytic approach is to assume all ground NVR remains on the surface throughout the flight phase of the mission. This is considered a conservative approach since the probability that molecules re-emitted on orbit will condense on another surface is considerably lower than the probability that particles excited during launch will deposit on another surface. For some surfaces that are very clean at launch and must remain so in orbit, NVR migration may warrant further investigation.

Data relating NVR_0 to typically used cleaning techniques is not currently available. The determination of this level is therefore the responsibility of the experiment manufacturer. If solvent type cleaning activities are not planned at the launch site, then NVR_0 becomes the NVR level present at the time the experiment arrives at the site. If the experiment is sensitive to contamination, the designer should determine the level he can tolerate at launch, and work back to the level necessary at arrival. In general, a precision cleaned surface may be considered to be at the highest NVR cleanliness level specified in Reference C-9, Level A, equal to or less than 1 mg/0.1m^2. This should be a conservatively high value for NVR_0 for a surface cleaned explicitly for NVR removal. After NVR_0 is established, the level of NVR present at launch may be calculated by summing the deposition rates for each facility multiplied by the time spent in each of the facilities.

Example 2.3.4.2:	Predict the NVR level present at launch for a surface exposed during the entire ground flow experienced by the IECM on SL-1 (STS-9). Assume an initial NVR level of 1.0 mg/0.1m^2.
Solution:	NVR collection was not conducted during the SL-1 ground flow. High recorded values for facilities in other recent flows will be used. From the tables of Section 2.2.2.2, NVR levels for the O and C Building, OPF and PCR are 0.5 mg/0.1m^2/month, 0.31 mg/0.1m^2/month, and 0.7 mg/0.1m^2/month, respectively (Canister not considered since exposure time is normally very short). From Reference C-14, the IECM was exposed in the O and C Building for approximately four months, and the duration in the OPF and PCR was a total of one month (one half month will be assumed for each facility). Applying Equation 2.2-4 gives a total NVR level at launch of 3.5 mg/0.1m^2.

2.2.5 Ground Contamination Preventative Techniques

Once it has been determined that an experiment is sensitive to contamination, and an evaluation of the data available indicates that the ground environment may degrade the performance of the instrument on orbit, preventive measures for use on the ground should be considered.

Reference C-9 presents the contamination control requirements for the STS program. This includes the minimum requirements a payload must meet to prevent contamination of other payloads, and provisions that may be used to reduce the contamination of sensitive payloads. Contamination reduction techniques applicable to ground activities include the development of a contamination control plan in order to identify operations to prevent unforeseen contamination prior to launch; design of equipment to allow contamination removal; selection of controlled environments for manufacture, assembly, test, shipment and integration, should they be deemed necessary; contamination detection and monitoring provisions; packaging provisions to protect hardware when access is not necessary; evaluation of subcontractor operations and component delivery cleanliness; control of fluids used for ground servicing and tests; identification of cleaning processes compatible with the hardware design and materials; and training of personnel in contamination control techniques. More detailed information on ground contamination control is available in several other documents, including Reference C-49 (design, cleaning, cleaning agents, cleanrooms, microbes, packaging procedures, packaging materials, packaging verification, cleanliness maintenance, and personnel activities), Reference C-50 (cleanrooms), Reference C-4 (cleanrooms), and Reference C-51 (inspections, cleanrooms, personnel, tools, garments, cleaning, packaging, fabrication, maintenance, and fluids).

In addition to the above traditional control techniques, some STS payloads currently plan to use localized HEPA filtered air or a dry N_2 purge, and shrouds or covers to be removed shortly before launch.

2.2.6 Key Ground Contamination Technical Personnel

The following personnel are currently active in the characterization and assessment of ground contamination related to spacecraft, and particularly related to the STS/SL:

	Organization	Telephone
Don Bartelson	LSOC	(303) 867-0960
Brent Wenkstern	MDAC	(305) 867-4328
Virginia Whitehead	KSC	
J. M. Ragusa	KSC	
Lubert J. Leger	JSC	(713) 483-2059
Steve Jacobs	JSC	(713) 483-3561
Edgar R. Miller (IECM)	MSFC	(205) 544-7752
R. C. Linton (PSA)	MSFC	(205) 544-2526
Jack Barengoltz	JPL	(818) 354-2516

	Organization	Telephone
Carl R. Maag	JPL	(818) 354-6453
Lyle E. Bareiss	Martin Marietta	(303) 977-8713
John H. Raab	Martin Marietta	(303) 977-1878
Larry Dell	Martin Marietta	(303) 977-1881
Gene N. Borson	Aerospace Corp.	(213) 648-6943
Larry H. Rachal	Aerospace Corp.	(213) 648-7646
Capt. N. S. Fulks	USAF/VAFB	

2.2.7 Ground Contamination References

C-1. Ragusa, J. M., "KSC Shuttle Ground Turnaround Evaluation", NASA KSC, in C-40, February 1983.

C-2. Hetrick, M. A., "Trip Report: VLS Contamination Workshop", I. D. 6496-34-84056, Martin Marietta, 2 November 1984.

C-3. Faenza, G. R., "KSC Cargo Facility Contamination Control Plan", K-STSM-14.2.1, MDTSCO for NASA KSC, 15 December 1983.

C-4. "Federal Standard Clean Room and Work Station Requirements, Controlled Environment", Fed. Std. No. 209B, General Services Administration, 24 April 1973.

C-5. Wenkstern, B., "SPIF Certification", MDTSCO, 11 October 1984.

C-6. "Military Standard Product Cleanliness Levels and Contamination Control Program", MIL-STD-1246A, 18 August 1967.

C-7. Bartelson, D. W., "Shuttle Facility Contamination Control Implementation Plan OPF PCR", KVT-PL-0025, Rockwell International for NASA KSC, January 1984.

C-8. Faenza, G. R., "Cargo Facility Contamination Control Implementation Plan", KCI-HB-5340.1, MDTSCO for NASA KSC, 15 December 1983.

C-9. "Specification Contamination Control Requirements for the Space Shuttle Program", SN-C-0005A, NASA JSC, January 1982.

C-10. Bartelson, D. W., "Contamination and Environmental Data During Payload Bay-Cargo Processing of STS-41B in the PCR", KVT-PL-0025, Addendum PCR-ED-STS 41B, Lockheed SOC, 17 April 1984.

C-11. Linton, R. C., E. R. Miller and M. Susko, "Passive Optical Sample Assembly (POSA): Final Report", NASA TM-82466, MSFC, August 1981.

C-12. Linton, R. C. and E. R. Miller, "Passive Optical Sample Assembly (POSA) for STS-1--Quick Look Report", NASA TM-82421, MSFC, 1 June 1981.

C-13. Van Dyke, D. M., "Measurement of Particulate Fallout Rates for Designated Martin Marietta Cleanrooms", presentation, Martin Marietta, April 1984.

C-14. Miller, E. R., ed., "Induced Environmental Contamination Monitor Preliminary Results from the Spacelab 1 Flight", NASA TM-86461, MSFC, August 1984.

C-15. Norris, R. E., "Shuttle-Cargo Integration Working Group September 26, 1985 Meeting Minutes", NASA KSC.

C-16. Borson, E. N., "Status of STS Contamination Control", presentation, Aerospace Corporation, 20 January 1982.

C-17. "STS-2 Test Milestone Report for the KSC Ground Operations Phase", 6555th Aerospace Test Group (ASTG), 1 March 1982.

C-18. Miller, E. R., "SL-1/Induced Environmental Contamination Monitor Very Quick Look Report", NASA MSFC, March 1984.

C-19. "Minutes of Contamination TIM Held at ELS, 3 December 1985".

C-20. Marrs, K. S., "Minutes of Contamination Working Group Meeting at ELS on Sept. 11 and 12, 1984", 2-1616-KJM-824, Boeing Aerospace Company, 28 September 1984.

C-21. Whitehead, V., "OPF Environment STS-3 Processing 11/23/81-1/25/82: Presentation to OPF Contamination Control PRB", NASA KSC, 18 March 1982.

C-22. Duncan, T. W. "OPF Contamination Control PRB", Memo No. 7150-82-054, Martin Marietta, 24 March 1984.

C-23. Whitehead, V. and T. Harris, "Relationship Between Air and Surface Cleanliness Classes at Kennedy Space Center Processing Facilities", NASA KSC and MDTSCO, October 1982.

C-24. Duncan, T. W., "OPF Environment During STS-2/OSTA-1 Processing", Memo No. 7150-81-515, Martin Marietta, 7 October 1981.

C-25. Miller, E. R., "STS-2 Induced Environment Contamination Monitor (IECM) - Quick Look Report", NASA TM-82457, MSFC, January 1982.

C-26. Miller, E. R. and J. A. Fountain, "STS-3 Induced Environment Contamination Monitor (IECM) - Quick Look Report", NASA TM-82489, MSFC, June 1982.

C-27. Leger, L. J., H. K. F. Ehlers, S. Jacobs, and E. R. Miller, "Space Shuttle Preliminary Contamination Assessment from STS-1 and STS-2", N82 25307 in NASA CP 2229, NASA, May 1982.

C-28. Rhodes, R., "OPF Contamination Control PRB: Presentation to PRB", NASA KSC, 18 March 1982.

C-29. Bartelson, D. W. "Cargo Community Familiarization With KSC-PLB/Cargo Integration Operations—Contamination Concerns", Lockheed SOC, 23 May 1984.

C-30. Norris, R. E., "1st Annual ELS Contamination Conference", NASA KSC, June 1985.

C-31. Norris, R. E., "Shuttle-Cargo Integration Working Group July 25, 1985 Meeting Minutes", NASA KSC.

C-32. Miller, E. R., "STS-2,-3,-4 Induced Environmental Contamination Monitor (IECM)", NASA TM-82524, MSFC, February 1983.

C-33. Norris, R. E., "Shuttle-Cargo Integration Working Group December 5, 1985 Meeting Minutes", NASA KSC.

C-34. Norris, R. E., "Shuttle-Cargo Integration Working Group April 5, 1985 Meeting Minutes", NASA KSC.

C-35. Blume, M. L., "Contamination and Environmental Data Recorded in the VPF and PCR During STS-7 Cargo Processing", MDTSCO MDC Y0262.

C-36. Norris, R. E., "Shuttle-Cargo Integration Working Group February 7, 1985 Meeting Minutes", NASA KSC.

C-37. Norris, R. E., "Shuttle-Cargo Integration Working Group May 9, 1985 Meeting Minutes", NASA KSC.

C-38. Norris, R. E., "Shuttle-Cargo Integration Working Group August 29, 1985 Meeting Minutes", NASA KSC.

C-39. Raab, J. H., "Quantification of Shuttle Orbiter Payload Bay Cleanliness Levels", MCR-86-2004, Martin Marietta, January 1986.

C-40. Tanner, J. G., and T. Wilkerson, eds., "The Shuttle Environment Workshop", Systematics General Corporation, February 1983.

C-41. Dell, L., J. H. Raab and M. A. Hetrick, "Investigation of Tape Lift Method in Quantifying Particle Contamination", MCR-85-2017, Martin Marietta, January 1986.

C-42. Raab, J. H., "Spacecraft/Shuttle Orbiter Ground and Ascent Contamination Analysis", MCR-84-2022, Martin Marietta, February 1985.

C-43. Kelly, J. G., "Measurement of Particle Contamination", AIAA-85-7003, in AIAA Environment and Operations II Conference: A Collection of Technical Papers, November 1985.

C-44. Hamberg, O. and E. M. Shon, "Particle Size Distribution on Surfaces in Clean Rooms", Aerospace Corporation, published by Institute of Environmental Science, May 1984.

C-45. Hamberg, O., "Particulate Fallout Predictions for Clean Rooms", Aerospace Corporation.

C-46. Yound, R. P., "Degradation of Low Scatter Mirrors by Particle Contamination", AEDC-TR-74-109, ARO, Inc., January 1975.

C-47. Reul, R. P., C. E. Hilbers, and E. Goller, "A Forecasting Technique for Accumulated Particulate Contamination of Spacecraft Assemblies", TRW Technical Report 82078A, TRW Corp., 30 October 1970.

C-48. Bareiss, L. E., F. J. Jarossy, and J. C. Pizzicaroli, "The Shuttle/Payload Contamination Evaluation Program: The SPACE II User's Manual", MCR-80-593, Martin Marietta, September 30, 1980.

C-49. "Contamination Control Handbook", NASA SP-5076, Sandia Labs for NASA MSFC, 1969.

C-50. Useller, J. W., "Clean Room Technology", NASA SP-5074, Lewis Research Center, 1969.

C-51. "Contamination Control Requirements Manual", JSCM 5322 Rev. A, NASA JSC, December 1974.

2.3 FLIGHT CONTAMINATION

Flight environments are those occurring between the times of launch and touchdown, thus including ascent, orbital and descent mission phases. The orbital phase is defined as the portion of the mission during which the Orbiter PLB doors are open, while the phases in which the doors are closed are considered either ascent or descent. This section provides information on the particulate and molecular contamination environments occurring during all flight mission phases.

2.3.1 Flight Introduction

Included in this section are a summary of measurement techniques used to collect flight data, available data on the flight environments and their effects, summaries of unpublished, needed and planned data collections not yet available, summaries of available models and tools for the prediction of contamination levels and effects, flight contamination preventive techniques, and key technical personnel in the field. This information is presented in two major sections: flight particulate contamination, and flight molecular contamination.

2.3.1.1 Environment Overview

The flight contamination environment includes particles and molecules transported from a contamination source to a sensitive surface during ascent, orbital operations, and descent. For surfaces observing through a field-of-view while on orbit, the contamination environment also includes molecules and particles within and passing through the field-of-view of the surface. In addition, since contamination effects are usually most noticeable during orbital operation, ground contamination remaining on a surface when orbit is achieved may be considered part of the flight contamination environment (from an effects standpoint).

2.3.1.1.1 Flight Particle Environment Overview

The initial flight particle environment of interest occurs doing launch and ascent. The vibroacoustic environment is very energetic during launch, particularly during SRB firing for the first phase of flight and the transonic phase (aeronoise through the PLB vents). It is believed that this vibroacoustic environment provides the energy necessary to excite some of the particles present the PLB at launch. The venting of the PLB volume through vents in the PLB lower walls and the acceleration of the Orbiter (which approaches 3 g's at the end of the SRB and main engine burns) provide transport mechanisms to carry excited particles from their original launch locations to other surfaces in the PLB. Thus, during launch a general redistribution of PLB particles is felt to occur.

While on orbit, mechanical and acceleration activities, including Orbital Maneuvering System (OMS), Primary Reaction Control System (PRCS) and Vernier Reaction Control System (VRCS) engine firings, Remote Manipulator System (RMS) and PLB door movements, crew activities, and other motion causing events can continue to excite particles throughout the mission. Such particles have been observed by crews on each Shuttle flight. In addition, other activities which occur throughout any given mission, such as engine firings and water dumps have been observed to generate additional particles.

Descent operations provide another opportunity for particles to be redistributed or ingested. While the vibroacoustic environment during descent is not as severe as the ascent environment, the repressurization of the PLB provides an additional particle transport mechanism for this phase. During descent, the deceleration vector is essentially at right angles to the acceleration vector during ascent. The descent orientation corresponds to an horizontal ground configuration, while the ascent orientation corresponds to a vertical to upside down ground configuration. This means that different PLB and payload surfaces would be expected to collect particles during ascent and descent.

2.3.1.1.2 Flight Molecular Environment Overview

Non-metallic materials will outgas molecular species such as plasticizers and unreacted monomers from plastics when exposed to the vacuum of space. As they outgas, such species have the potential to chemically and/or photochemically react with other absorbed species. Oxidative reactions are known to occur in low earth orbit (LEO) and can produce new and different molecules. Leaks from cooling systems may release coolants such as Freons or Coolanol. During ascent HCl from the solid rockets can be carried to altitude and, during descent, nitrogen products from the reaction control system and auxiliary power system may find their way into the cargo bay. Clearly the payloads and the Shuttle itself have the potential for contaminating the PLB environment with molecular species.

2.3.1.2 Flight Contamination Measurement Techniques

Several contamination monitors and instruments from which contamination levels may be inferred have been flown on the Space Shuttle Orbiter. These include the POSA on STS-1, the IECM on STS-2, STS-3, STS-4 and STS-9, the Contamination Monitor Package (CMP) (non-contamination role on STS-8), Microabrasion Foil Experiment (MFE), the Shuttle/Spacelab Induced Atmosphere (SIA) camera and forward PLB T.V. cameras on STS-3, the Air Force Geophysics Laboratory (AFGL) CIRRIS-1A infrared telescope and mass spectrometer on STS-4, and the Shuttle Pallet Satellite (SPAS) on STS-7 and STS-11. Other instruments for which data is not completely available, is classified, or which have not yet flown are the AFGL Particle Analysis Cameras for Shuttle (PACS), the Air Force/JPL Interim Operational Contamination Monitor (IOCM), and the infrared telescope (IRT) of SL-2.

2.3.1.2.1 Induced Environment Contamination Monitor (IECM)

As a result of concern for possible contamination from the environment of Shuttle which might place limitations on experiment measurements, the IECM was used to provide measurements during ascent, on-orbit, and descent in order to determine the actual environment. The IECM, designed and integrated by MSFC, is comprised of ten instruments (Ref. D-1):

(1) The Humidity Monitor (IECM01) is used to measure the humidity/temperature profile of the environment within the cargo bay of the Shuttle. It is an off-the-shelf version of the Model 2000 built by Thunder Scientific Corporation, Albuquerque, New Mexico, with minor modifications. It measures relative humidity from 0 to 100 percent with an accuracy of ± 4 percent over a temperature range of 0 to 70°C. The temperature measurement (0 to 100°C) is made with a thermistor located within the humidity sensor mounting. The sensing element detector is a Brady array which can sense a 0.1 percent change in humidity, but hysteresis and thermal compensation limits the overall accuracy to ± 4 percent. The sensor is mounted in the manifold of the Air Sampler (IECM03, discussed below).

(2) The Dew Point Hygrometer (IECM02) measures the dew point of the air surrounding the IECM. Measurements are made prior to launch and as long as the vehicle is within the Earth's atmosphere, including ascent, reentry, and landing. It was built by EG and C, Inc. for the Skylab program and is a retrofit from that program. Measurements are made over a temperature range of -6.7°C to 26.7°C with an accuracy of 0.5°C. The operating temperature range (dew point) is -6.66°C to 26.66°C with a non-operating range of -28.0°C to 70.1°C. The time for a 63 percent response to a 11.7°C step change in dew point temperature is nominally 10 sec. The sensor is mounted within the air manifold of the Air Sampler (IECM03).

(3) The Air Sampler (IECM03) measures the gaseous contaminants in the cargo bay during ground based operations, ascent, and descent. It was built by Spacecraft, Inc., Huntsville, Alabama. It consists of five stainless steel sample bottles attached to a pumping manifold used to draw samples into the collection bottles. The bottles are appropriately packed with adsorbents Tanax GC and Spherocarb for molecular species, platelets of Ag_2O for reactive materials such as HCl (expected on ascent), and platelets of ruthenium compounds for NOx/NH_3 (expected during descent). At approximately 0.25 atm the pumping is no longer effective; therefore, the ascent bottles were evacuated before flight, and the descent bottle was opened in-orbit for evacuation. Chemical analysis was carried out post-flight in ground-based laboratories.

(4) The Cascade Impactor (IECM04) is a series of quartz crystal microbalances (QCMs), used to measure the volumetric particle content of the PLB during ground, ascent and descent phases of flight. The QCMs are similar to those described for IECM07, without temperature control. For the first stage, the sensing crystal is exposed to the ambient environment while the reference crystal is sealed and remains unexposed. This stage is used to detect molecular contaminants that could alter the results from the other stages. The remaining three stages each consist of a nozzle directing flow toward the sensing crystal, and a spacer to shield the reference crystal. The nozzle of each stage is sized such that particles smaller than a given size will be carried around the reference crystal by viscous drag forces. The optimal size at which 50 percent of the incident particles strike the crystal surface are 5 µm, 0.816 µm and 0.248 µm for stages 2, 3 and 4, respectively. These correspond to effective collection ranges of >5 µm, 0.8 µm to 5 µm, and 0.25 µm to 0.8 µm for the three stages. The sensing crystals are coated with a layer of Apiezon grease to cause impacting particles to stick on the crystal. A constant volume pump

(250 ml/min) is used to induce flow through the instrument. The pump is turned off when the air pressure in the PLB decreases sufficiently to reduce particle flow.

Recent analyses by one of the Cascade Impactor investigators (Ref. D-2) have raised doubts about the actual flow volume induced in the instrument and the instrument's geometrical design. An accurate knowledge of the flow volume is needed to translate the mass collected to a volumetric particle count, and the geometry of the instrument may result in preferential counting of particles originating within the IECM, particularly for particles larger than 10 µm. Thus at this time, Cascade Impactor results must be used with caution, and possibly only in a comparative manner.

(5) The Passive Sample Array (IECM05) provides an array of optical witness samples to be exposed to the natural and induced environment of the Shuttle cargo bay and the Long Duration Exposure Facility (LDEF). The samples are measured in the laboratory prior to integration. Control samples are included in these measurements and are then stored in a controlled, clean environment. Following retrieval of samples, whether during preflight activities or post-flight, the measurements are repeated and the analysis is based on changes from the previous measurements. Each sample holder (stainless steel) provides six 2.54 cm depressions to hold the samples. Trays hold up to eight such sample holders. The array is flush-mounted at the top of the IECM with virtually no shadowing of any sample in the array.

(6) The Optical Effects Module (OEM, IECM06) measures degradation of optical window materials in transmittance and scattering. Optical property changes due to deposition of particles and molecular films are discriminately measured utilizing an integrated scattered light measurement in conjunction with direct, self-calibrating transmission measurements. It consists of a light source, focusing and collecting optics, a rotatable sample carousel, and detectors. It is nominally monochromatic at 253.65 nm, determined by choice of the light source and spectral sensitivity of the detectors. The instrument was developed by Advanced Kinetics, Inc., Costa Mesa, California.

The samples are 2.54 cm diameter, 3 mm thick optical flats of which only three samples are exposed to the external environment at one time.

A nearly collimated beam of light from the source is uniquely differentiated from any background illumination by passing through a Bulova tuning fork chopper at 200 Hz. One sample position on the carousel is left blank to provide the I_0 calibration reference for all measurements. When the carousel is in a filled position, the intensity I_n is ratioed to I_0 as a measure of the transmittance.

The optical alignment is designed to reject the specular component of reflection from the sample (with a half-cone angle of 10 deg). If diffuse reflection or scattering occurs on a sample, the focusing mirror is positioned to collect this type flux and direct it to a second photomultiplier above the transmission detector for a measurement of scattered intensity, I_s, similarly ratioed to I_0.

(7) The Temperature-Controlled Quartz Crystal Microbalance (TQCM, IECM07) detects the adsorption or desorption of molecular contamination in the cargo bay as a function of temperature. It was developed by Faraday Laboratories, Inc., LaJolla, California, and has five identical sensor heads. Each sensor consists of two identical crystals. The sensing crystal is directly exposed to the environment, and its reference crystal is mounted directly behind the sensing crystal and is, therefore, shielded from direct line-of-sight to the molecular environment. They operate at 15 MHz with coatings applied to the crystals in such a manner as to change the operating frequency of the sensing crystal to be approximately 1 KHz less than the reference crystal. The two frequencies are mixed electronically, and any deposition which lowers the frequency of the sensing crystal is measured as an increase in the beat frequency.

The sensors have a sensitivity of 1.56×10^{-9} g/cm^2 Hz. The crystal has a finite limit to the amount of material it can detect. At approximately 1×10^{-4} g/cm^2 of deposition, the mass becomes large enough to damp the crystal from oscillating. The crystal is then heated to desorb contamination and return it to active oscillation. Contaminants that may be photo-polymerized and cannot be removed by heating are determined by observing the permanent frequency shift after volatiles are removed by heating.

The sensors faced the +X (forward), -Z (out of bay) -X (aft), +Y (right) and -Y (left) directions of the Orbiter axes.

(8) The Cryogenic Quartz Crystal Microbalance (CQCM, IECM08) provides a record of adsorption and desorption of molecular contamination in the cargo bay with the special objective of measuring water vapor. This is accomplished by the passive radiative design which causes the detector crystal to cool to the temperature at which water vapor will condense. It was also developed by Faraday Laboratories and uses the technology described for IECM07.

(9) The Camera-Photometer (IECM09) measures the induced contamination in the form of individual particles and general background. Two automated systems are placed on-board the IECM. They were developed by Epsilon Laboratories, Inc., Bedford, Massachusetts. Each is housed within a canister and pressurized to 1 atm with a standard atmospheric mixture of gases. An initial relative humidity of 20 percent is utilized to prevent film damage.

Observations are made by the camera through a quartz window, and a multivaned baffle system is utilized to prevent stray light from raising the background intensity. The baffle rejects scattered light to levels below 10^{-14} solar brightness (B_0) for solar angles greater than 60 degrees from the optical axis. The camera is a 16 mm Model H-16SB Bolex movie camera with a nominal film capacity of 4000 frames. The camera lens is an 18 mm f/0.9 with a nominal field-of-view of 20° half-angle (limited by the baffle to 10°). The camera operates at a rate of 24 frames/ hour.

Operating in conjunction with the camera is an integrating photometer which monitors continuously the background brightness over a dynamic range of 10^3 centered at 10^{-13} B_0. The photometer also functions to control exposure time by terminating any exposure which exceeds a predetermined brightness level. This value is adjustable to correspond to film exposure requirements.

(10) The Mass Spectrometer (IECM10) collects data from which molecular column density and molecular return flux may be inferred. The instrument is designed to measure collimated flux with a view angle of 0.1 sr. It is a quadrapole instrument and was developed by the Space Research Institute of the University of Michigan to measure the pressure (density) of all gases separated from 1 to 150 amu. In-flight calibration is performed by a gas release system which emits a known flux of isotopically labeled water and neon into the collimated view of the spectrometer; the backscattered flux is then measured. The calibration measurements also provide the basis for evaluating differential scattering cross-sections for 8 km/sec collisions (approximate orbital velocity).

2.3.1.2.2 Other Contamination Monitors

Several other contamination monitors or devices providing information on the flight contamination environment have been flown on various STS missions. These include:

(1) The Contamination Monitor Package of STS-3 was developed by JPL under funding from the USAF. It consists of four TQCMs and two passively controlled witness samples. The CMP witness plates were two MgF_2 coated mirrors provided by different vendors and placed on the -Z surface of the CMP. One-half of each sample was shielded from exposure to the sun while on orbit. Similar samples were exposed to the ground flow environments, but not flown in order to act as controls. Pre- and post-flight reflectance measurements were made on the flight and control samples (Ref. D-3).

(2) The Shuttle Pallet Satellite Mass Spectrometer is a Maltauch-Herzog geometry mass spectrometer and was flown on STS-7 and STS-41B. It is an electrostatic magnetic double focusing spectrometer with a four-fold ion detector system: 1) a monitor for total ion current (TIC) passing through the analyzer; 2) two ion counting multipliers, one for masses below 10 amu and one for those from 10 to 80 amu; and 3) a dc measuring electrometer for masses greater than 10 amu (HM electrometer). The mass spectrometer took eight readings per second from each detector and was programmed to measure periodically a few selected masses. Once every 15 minutes a continuous spectrum was taken which took approximately 4 minutes (Ref. D-4).

(3) The AFGL Mass Spectrometer was flown on STS-4 in conjunction with CIRRIS. The instrument operated in two modes under ground control: a positive ion mode, and a neutral gas mode. Both modes included both unbiased and retarding voltage operations to differentiate between low energy (Shuttle related) and high energy (ambient) gases. The instrument orientation was horizontal along the Orbiter right wing (Ref. D-5).

(4) CIRRIS is an cryogenic infrared telescope that was flown on STS-4 by the AFGL and is planned for reflight. Although primarily intended for remote sensing of infrared sources, the telescope additionally can detect near-field particles within its 1° field-of-view.

(5) The POSA was a set of five optical witness samples and three statically charged Teflon sheets (electrets). The POSA samples were mounted on the starboard rail of the STS-1 Development Flight Instrumentation (DFI) pallet in the Orbiter PLB. The POSA samples were installed in the PLB approximately 1 month prior to launch, and remained exposed throughout the mission, landing, and return to the OPF after ferry flight. A second similar POSA was installed at the landing site prior to ferry flight to assess the relative impact of the post-landing phase compared to the prelaunch and flight phases. Evaluations of the samples after return to MSFC included photography, transmittance and reflectance measurements, particle counts, x-ray microprobe analyses of the electrets, and scanning electron microscopy (SEM) of the optical samples (Ref. D-6).

(6) The Microabrasion Foil Experiment was an experiment flown on STS-3. The purpose of the experiment was to detect orbital micrometeoroids. It consisted of a $0.4m^2$ capture cell array of which the upper surface was 5 μm thick aluminum foil. After flight, the capture cell was examined for micrometeoroid penetrations, and the presence of deposited particles in the foil, indicative of low velocity impacts, was detected. The particles were subsequently examined by energy-dispersive x-ray analysis (Ref. D-7).

(7) The Shuttle forward PLB Low Light Level (LLL) T.V. Cameras were used to gather images of particles generated in the PLB during STS-3. Videotapes from these cameras were then evaluated by JPL to determine particle sizes, sources and velocities (Ref. D-8).

(8) The Shuttle Induced Atmosphere (SIA) camera flown on STS-3 is a camera-photometer originally used for the Skylab program. It consists of a 16mm camera and a photometer mounted on a single axis gimbal to allow front to back scans. The instrument has a 6° field-of-view and a series of spectral filters. The instrument was used to determine optical emissions and reflections from Orbiter surfaces, the Orbiter atmosphere, and astronomical objects (Ref. D-9).

(9) The Spacelab Small Helium-Cooled Infrared Telescope (IRT) was flown on STS-51F/SL-2 (Ref. D-10). The instrument consists of a cooled (8 K) highly baffled Herschelian optical system with an off-axis 15-cm diameter, f/4 primary mirror. The telescope rotates in a plane about the Shuttle X axis. It scans at 6°/sec over a 90° arc of the sky. The focal plane of the telescope contains ten detectors; nine cover the spectral region from 4 to 120 μm in four broad non-overlapping bands (4.5-8.5, 9-14, 18-30 and 70-120 μ), and one with narrow band response at the 6-7 μm H_2O band. The detectors are masked with a single band to modulate the signal, to allow discrimination between near-field (particle) IR sources and stellar sources. The instrument experienced some difficulties in flight, including sunshade insulation in its field-of-view and saturation of several channels, making unambiguous interpretations difficult.

(10) The Particle Analysis Cameras for Shuttle (PACS) was flown on STS-61C. The instrument consist of two 35 mm cameras and a strobe. The system operated every 120 seconds, generating stereo images of particles in the camera field-of-view (Ref. D-11)

2.3.1.3 Flight Introduction References

D-1. Miller, E. R. and R. Decher, "An Induced Environmental Contamination Monitor for the Space Shuttle", NASA TM-78193, MSFC, August 1978.

D-2. Miller, E. R., ed., "Induced Environmental Contamination Monitor Preliminary Results from the Spacelab 1 Flight", NASA TM-86461, MSFC, August 1984.

D-3. Triolo, J., R. Kruger, R. McIntosh, C. Maag and P. A. Porzio, "Results from a "Small Box" Real-Time Molecular Contamination Monitor on STS-3", AIAA-83-0251, April 1984.

D-4. Wulf, E. and U. von Zahn, "Behaviour of Contaminant Gases Emitted by the Space Shuttle and the Manned Maneuvering Unit", Physikalisches Institute, Universetat Bonn, 2 August 1984.

D-5. Narcisi, R., E. Trzcinski, G. Federico, L. Wlodya and D. Delorey, "The Gaseous and Plasma Environment Around Space Shuttle", AIAA-83-2659, AFGL and Boston College, published by AIAA, November 1983.

D-6. Linton, R. C. and E. R. Miller, "Passive Optical Sample Assembly (POSA) for STS-1--Quick Look Report", NASA TM-82421, MSFC, 1 June 1981.

D-7. Dixon, D. G., W. C. Carey and J. A. M. McDonnell, "Contamination By Fibers on Space Shuttle Flight OSS-1 Microabrasion Foil Experiment", Journal of Spacecraft, Vol. 21, No. 4, April 1984, University of Kent, published by AIAA.

D-8. Barengoltz, J., F. Kuydendall, and C. Maag, "The Particle Environment of STS-3 as Observed by the Cargo Bay Television System", JPL, 25 October 1982.

D-9. Weinberg, J. L., F. Giovane, D. W. Schuerman and R. C. Hohn, "OSS-1/STS-3 Shuttle Induced Atmosphere Experiment", University of Florida, published by NASA, October 1982.

D-10. Clifton, K. S. (ed), "Spacelab Mission 2 Experiment Descriptions - Second Edition", NASA TM-82477, January 1982.

D-11. Redd, F. J. and M. Ahmadjian,"Particle Analysis Camera for Shuttle (PACS): The First Hitchhiker (HHG-1)", AIAA-85-7005, AIAA Shuttle Environment and Operations II, November 1985.

2.3.2 Flight Particulate Environment

This section and its subsections present the available models and data on the particulate environment and its effects during ascent, orbit and descent mission phases. Also presented are descriptions of unpublished, needed and planned data, contamination prevention techniques, key technical personnel in the field, and references in which more detailed information may be found.

Section 2.2.2.1 presented the available data on the particle fallout environments found in STS/Spacelab integration facilities, and Section 2.2.4 described how available data could be used to predict cleanliness levels of PLB surfaces at the time of launch. While the Shuttle is being launched, however, the Orbiter and its payloads experience a significant dynamic environment, and it is likely that large fractions of particles present on surfaces will be agitated and migrate within the payload bay. Dynamic events, including thermal accommodations, can continue to excite particles throughout the mission. Other orbital events, such as flash evaporator operations, can generate particles. This redistribution and generation of particles and their effects are the subject of the following sections.

2.3.2.1 Available Flight Particle Data

This section includes the available data on both the flight particle environments, and particle effects. The majority of the effects data are presented in Section 2.3.2.1.2, Orbital Particle Data. Additional effects information may be found in Section 2.3.2.3, Flight Particle Analytical Tools.

2.3.2.1.1 Particle Data During Launch

The principle data for the launch particle environment has been collected using the Cascade Impactor of the IECM. As described in Section 2.3.1.2, this device consists of 3 stages, each made up of a sized nozzle and a QCM covered by Apiezon grease to collect particles. A constant volume air pump draws PLB air through the nozzles during Orbiter launch (and also during descent). The pump operates for approximately 2 minutes from liftoff, and is shut off once air pressure drops to the point where aerodynamic forces on PLB particles are significantly reduced. The mass collected by the stages of the Cascade Impactor may be converted to volumetric densities of particles within three size ranges. A fourth QCM monitors molecular deposition so that the effects of significant molecular collections may be removed from the particle data.

Recent analysis has indicated "that for all the IECM flights, the metered airflow may have been too high, and the geometrical configuration provided preferential measurements of particles inside the IECM, especially for particles larger than 10 μm" (Ref. E-1). With this qualification, Table 2.3-1 lists the maximum particle counts recorded during the launches of STS-2, STS-3, STS-4 and STS-9. The maximum counts during ascent usually occur approximately 1 minute after launch (Ref. E-2).

Reference E-3 indicates that the Cascade Impactor was operated for approximately 11 hours on the pad during STS-2 preparation prior to launch. This nearly saturated the QCMs with particles. Thus the collection efficiency during that launch may have been less than nominal.

Table 2.3-1 Cascade Impactor Launch Particle Densities (Ref E-1, E-2)

Particle Size, μm	STS Flight Particle Concentration ($\mu g/m^3$)			
	STS-2	STS-3	STS-4	STS-9
0.25 – 0.8	250	10	150	Below
0.8 – 5	500	10	400	Detection
>5	30	10	Not Funct	Limits

Additional data related to the particle environment during launch has been inferred from passive samples, including the POSA of STS-1, the IECM PSA of STS-2, STS-3, STS-4 and STS-9, and the MFE of STS-3. These are all surfaces that were installed in the Orbiter PLB prior to launch, and were unprotected from the time covers were removed before launch through launch, orbital operations, descent, landing, ferry flight and deintegration. In the case of the POSA and PSA samples, additional similar POSA sets were installed between landing and ferry flight so that the post-landing environment could be quantified and subtracted from the flight readings. Since the samples are exposed throughout the entire mission, conclusions about specific mission phases are difficult to draw. It is assumed, however, that the vibroacoustic and acceleration environment encountered during launch results in the most severe environment occurring at that time. A second severe environment is expected to occur during descent, due to deceleration and the repressurization of the PLB, which may disperse particles concentrated on the PLB vent filters. It is speculated that this environment is less severe than that encountered during launch. Data from the Cascade Impactor presented in Section 2.3.2.1.3 substantiates this conclusion. While accelerations during orbital operations have been observed to generate particles in the PLB area, the fact that the PLB doors are open leads to the conclusion that particle deposition while on orbit would not be significant.

While the launch environment is felt to be more severe then the descent environment, the orientation of surfaces is important in determining settling on any given surface. The surfaces of the PSA and POSA were installed in the Orbiter X-Y plane, or what would be a horizontal position for a horizontal Orbiter. The vertical launch of the Orbiter combined with a roll-over maneuver shortly after launch means that these surfaces are in a mostly vertical, and partially inverted (upside-down) position during ascent. Thus it is unlikely that many large particles would be collected on these surfaces during launch. During descent, however, the Orbiter assumes a generally horizontal orientation, which would be more conducive to particle settling on these surfaces. Given these limitations on the PSA and POSA results, Figure 2.3-1 presents the average particle size distributions collected in flight. These distributions have been corrected by the post-flight particle distributions collected for each mission (Ref. E-2, E-4, and E-5).

The MFE which flew on STS-3 was designed to record micrometeoroid impacts in a 5 μm thick aluminum foil sheet. The orientation of the 0.4 m² collector has not been confirmed, but an orientation parallel to that of the POSA and PSAs is most probable given the goal of the experiment. In post-flight examinations of the experiment, 6 μm diameter glass rods were found embedded in the foil (Ref. E-6). The most likely source for such fibers is the teflon coated glass fiber beta-cloth used extensively on the inner surfaces of the PLB. Experiments were conducted to attempt to reproduce the penetrations found on the flight experiments. Vibration of a foil sheet in the presence of glass fibers and bombardment of the foil surface by fibers in a 50 m/s air stream both produced effects similar to those observed from flight, while wiping the foil surface with fibers present pressed the fibers longitudinally into the foil surface, and air settling of fibers onto the surface produced no penetrations. Thus it is concluded that the penetrations occurred during the dynamic environments of launch and descent.

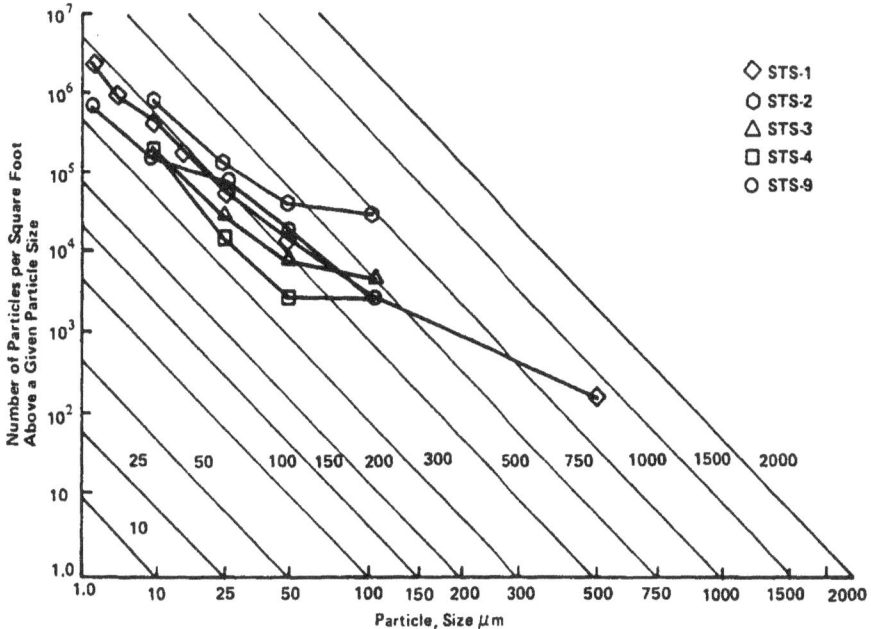

Figure 2.3-1
STS-1, STS-2, STS-3, STS-4 and STS-9 POSA and PSA Flight Particle Fallout (Ref E-2, E-4, E-5)

In summary, it may be stated that the launch particle environment data is limited and inconclusive. The ascent conditions may produce the most severe particle environments encountered by payloads during flight. Data to substantiate this hypothesis and quantify the levels encountered is not yet conclusive.

2.3.2.1.2 Orbital Particle Contamination Data

The primary particle environments of concern during orbital operations are released particles within the field-of-view of instruments, and the effects of both deposited and field-of-view particles on the instrument. Usable flight data is currently limited to photographic images of particles in the vicinity of the Orbiter PLB. Degradation of optical samples of the POSA, and PSA and OEM of the IECM, has generally been too small to provide usable effects data. Degradation has been measured in ground experiments, however, and these results may be applied to flight conditions.

Particle images have been collected by the IECM Camera-Photometer on STS-2, STS-3, STS-4 and STS-9 (SL-1), and by the PLB forward LLLTV cameras on STS-3. Usable images may be collected only during certain orbital periods. For the Camera-Photometer, this occurs when the Orbiter is sunlit and the camera and payload bay face a dark stellar or terrestrial background. The STS-3 T.V. camera images were collected in a tail-to-sun orientation.

Reference E-2 indicates that under ideal conditions, the Camera-Photometer could detect particles as small as 25 μm in the immediate vicinity of the Orbiter on STS-2, STS-3 and STS-4. Comparisons with calculations in Reference E-7 prior to film calibration indicate that the detection distance for this size particle is approximately 20 meters. Larger particles may be expected to be detected at greater distances. Reference E-4 indicates that the detectable particle limit was somewhat larger, at 28 μm for STS-9 (SL-1).

Figure 2.3-2 shows an averaged time history of the percentage of potential particle frames that actually detected particles based on STS-2, STS-3 and STS-4 images (Ref. E-2). All three flights observed particles in every photograph for which particle detection was possible taken during the first 7 hours of flight, and a general decline in the number of particles was observed with time. Figure 2.3-3 shows a similar plot for STS-9 (Ref. E-1). It shows similar high levels of particles during the early mission phase, and a general decay for the first 48 hours, although not as consistent as on earlier flights. After 48 hours, however, particle levels are consistently high. This may reflect the greater complexity and size of the SL-1 payload relative to the earlier flights. During water dumps, particle levels as high as 100/frame have been observed, with decay constants (1/e) of 5 minutes (Ref. E-2A). No increases in particle count have been observed for flash evaporator operations.

The data collected by the Orbiter PLB Low Light Level T.V. (LLLTV) camera used for the analysis reported in Reference E-8 are more difficult to interpret than the data from the Camera-Photometer, in that a single camera was used for the analysis. The Orbiter was in a tail to sun configuration when the images were recorded, backlighting the particles. With a single camera, however, sizes and distances may not be determined, so the analysts were forced to assume worst case (furthest aft) distances and particle sizes in most cases. Apparent particle sizes were determined from the video image sizes of the particles. A few particles were suffi-

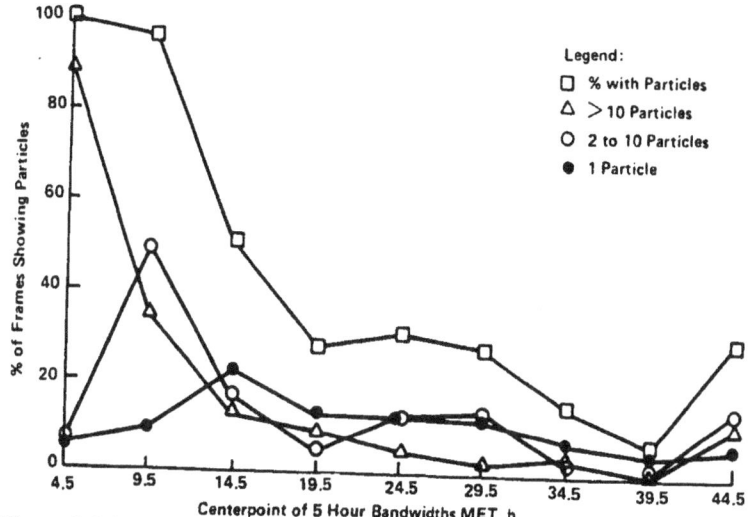

Figure 2.3-2
Percentage of Camera-Photometer Frames Showing Particles for STS-2, STS-3, and STS-4 (Ref E-2)

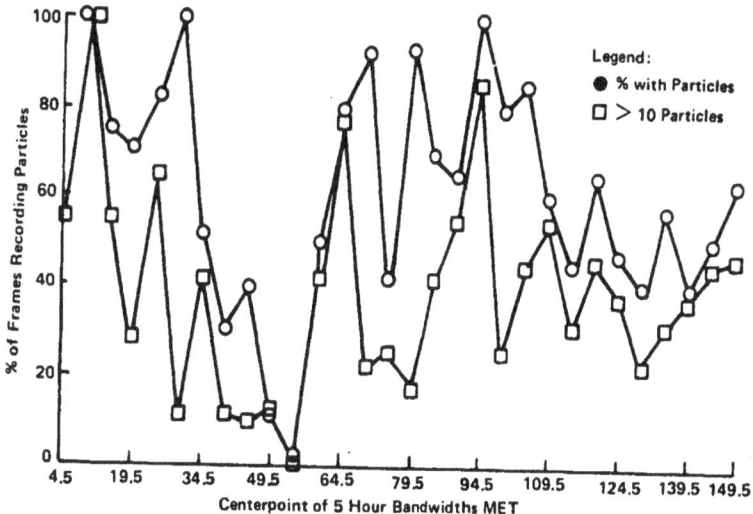

Figure 2.3-3
Percentage of STS-9 Camera-Photometer Frames Showing Particles (Ref E-1)

ciently small yet bright enough to show drag effects in their trajectories from which sizes and distances were determined by assuming the Orbiter trajectory was in the plane of the image and parallel to the starboard Orbiter wing. The velocity actually varied around this vector with orbit position. All particles for which this analysis could be performed were found to be smaller and closer than the worst case assumptions above. The assumptions necessary for the drag analysis also lead to considerable uncertanties.

Other difficulties in interpreting the LLLTV images make specific conclusions regarding particle numbers, size and locations difficult. The results do indicate a typical early flight particle event impact on a system with similar sensitivities to the LLLTV. For the event examined, as many as 60 particles as large as 5mm or larger were observed by the LLLTV in the 4° half angle about the Orbiter x-axis from the camera (Ref. E-8).

2.3.2.1.3 Particle Data During Descent

Descent particle data is very similar to the data for the ascent phase of flight. The POSA/PSA results cannot be separated into launch and descent phase effects, although, as was mentioned in Section 2.3.2.1.1, the orientation of these samples may make them better indicators of the descent phase than launch. The PSA and POSA results were presented in Figure 2.3-1.

In addition to the passive samples, the Cascade Impactor was also operated during descents of STS-2, STS-3, STS-4 and STS-9. The qualification of Cascade Impactor data presented in Section 2.3.1.2.1 applies equally to the descent phase. The data recorded by the device during descent is shown in Table 2.3-2 (Ref. E-2).

Table 2.3-2 Descent Cascade Impactor Results (Ref E-2)

Particle Size, μm	STS Flight Particle Concentration ($\mu g/m^3$)			
	STS-2	STS-3	STS-4	STS-9
0.25 – 0.8	125	10	Not Functioning	Not Published
0.8 – 5	250	10	10	
>5	10	10	20	

2.3.2.1.4 Particle Effects Data

Particle effects may generally be classed as obscuration effects, scattering effects and emission effects. Obscuration effects are those caused by the particle obscuring the surface of interest. In general, for obscuration effects the optical properties of the particle replace the properties of the obscured portion of the surface, degrading the surface performance accordingly. Scattering effects occur when particles affect the transmission properties of a surface or volume or increase diffuse reflectance. Particle emission effects are generally important only to infrared detectors.

Reference E-9 describes the results of experiments in which mirrors were contaminated with 3 μm and 20 μm aluminum oxide particles and fly ash. Obscuration ratio, surface reflectance and surface absorptance were measured for these contaminants. Figure 2.3-4 shows reflectance changes for 3 μm and 20 μm aluminum oxide particles as a function of obscuration ratio at wavelengths of 0.25, 0.3, 0.4, 0.65, 1.2 and 1.6 μm. Figure 2.3-5 shows similar changes for fly ash contamination. Figure 2.3-6 shows the change in solar absorptance for these same contaminants.

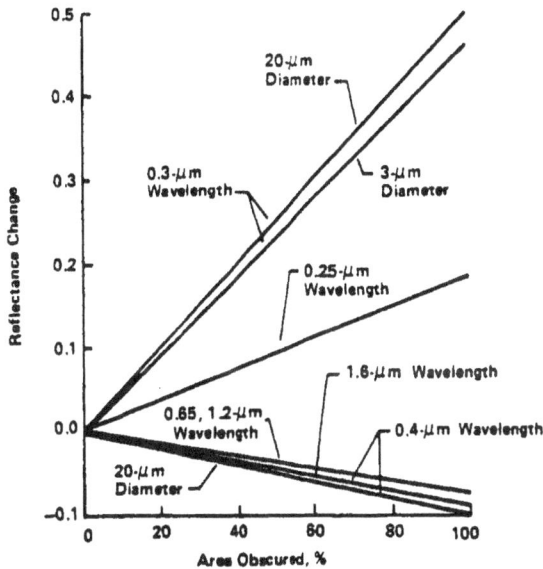

Figure 2.3-4
Reflectance Change Induced by 3 and 20 micron
Al_2O_3 Spherical Contamination (Ref E-9)

The most important observation from the three figures is the linearity of the reflectance and absorptance changes as functions of obscuration ratio. Section 2.3.2.3.2 will describe how this linearity may be used to translate predicted deposition levels to predicted degradations.

The fly ash data may also be used as a conservative measure of contamination effects, since it demonstrates reflectance degradation at all wavelengths from 0.25 μm to 1.6 μm, and the contaminant causes large changes in solar absorptivity.

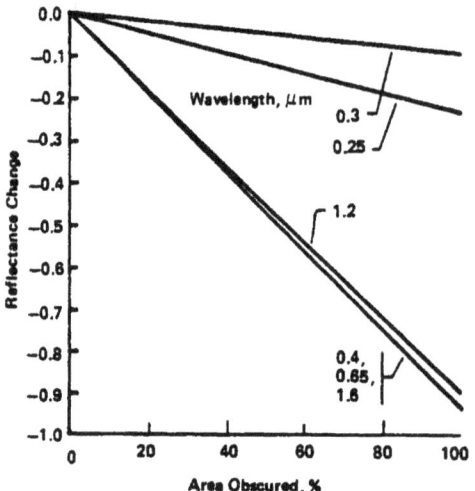

Figure 2.3-5
Reflectance Change Induced by Fly Ash Contamination (Ref E-9)

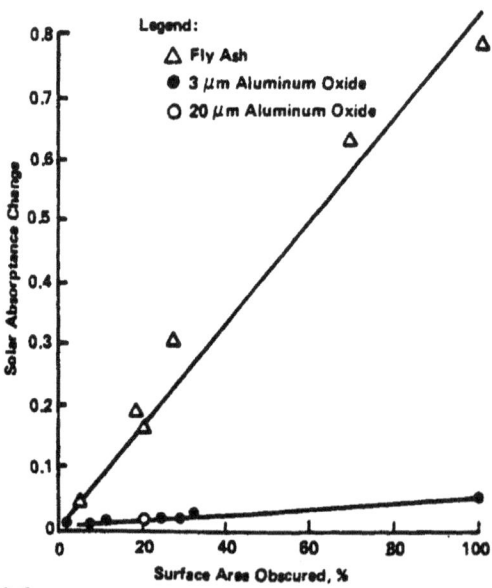

Figure 2.3-6
Solar Absorptance Change Versus Surface Area Obscured (Ref E-9)

The scattering effects of deposited particles have been reported in References E-10, E-11 and E-12. One measure of the effects of particle contamination is the bidirectional reflectance distribution function (BRDF), defined as the intensity of light reflected from a surface relative to incident intensity as a function of angle from the specular reflectance angle, normalized to 1 steradian. For near normal incidence,

$$BRDF = \frac{P_s}{P_i \Omega \cos \theta} \qquad 2.3\text{-}1$$

where

P_s = scattered light energy;
P_i = incident light energy;
Ω = solid angle of detector; and
θ = angle of detector from specular reflection (Ref. E-10).

Figure 2.3-7 shows the BRDF at 10.6 µm and 0.63 µm wavelengths measured on low scatter mirrors exposed vertically in a laboratory environment for 1176 hours (Ref. E-10). An attempt was made to count the particles present on the surface and calculate the surface obscuration ratio. Microscopic evaluation of the surface, however, revealed that the majority of the obscuration was caused by particles from 0.25 µm to 1.0 µm in size. These were too small and numerous to count accurately with the microscopic techniques used. An estimate of obscuration based on a small sample area and assumed 0.025 µm diameter circular particles was 2.3×10^{-5}.

Additional particle effects data could conceivably have been collected by the OEM of the IECM. In general, however, the effects observed by the OEM have not been significant enough to provide usable data.

2.3.2.2 Additional Flight Particle Data

2.3.2.2.1 Unpublished Flight Particle Data

One set of unpublished particle data was collected by JPL during a Shuttle-deployed solid upper stage firing. A foil specimen attached to the Orbiter Remote Manipulator System (RMS) was exposed to the firing engine while the PLB was turned away from the firing. Also, the IOCM described in Section 2.3.2.2.3, was flown on STS-51C (Department of Defense, (DOD) flight). The report of the data collected, however, is classified, and thus is considered unpublished from the point of view of this Handbook (Ref. E-13).

The IRT of SL-2 detected particles within its field-of-view, but very little data has yet been analyzed or published. This data may provide information on particle generation rates and particle temperatures.

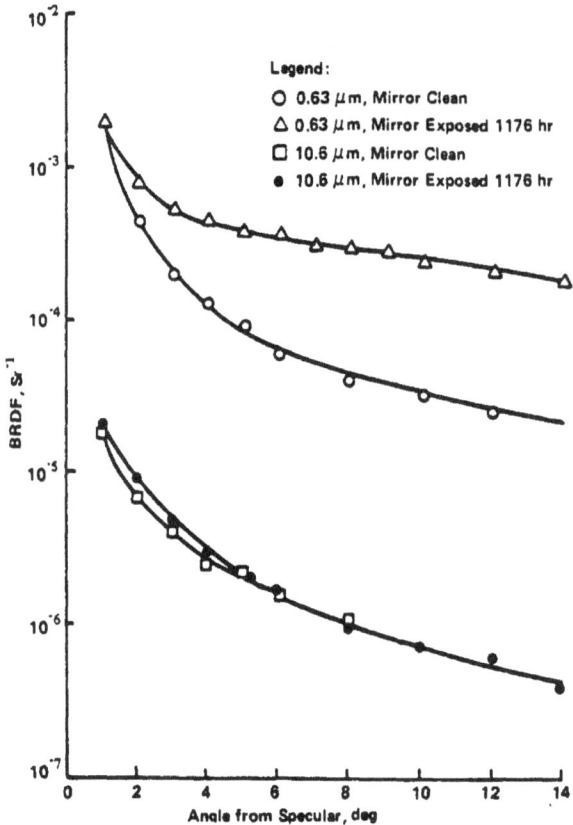

Figure 2.3-7
BRDF Before and After 1176-h Exposure, Mirror Position Vertical (Ref E-10)

More recent particle data was collected on STS-61C, when the AFGL/AFSD Particle Analysis Camera for Shuttle (PACS) was flown on a GSFC Hitchhiker payload carrier. The system collects photographic images using two cameras and a strobe operating every 120 seconds. It should have detected particles (as small as 25 μm in diameter) floating above the PLB.

In addition, some of the particle data collected by the IECM on SL-1 is being evaluated in greater detail by MSFC, and is planned for future publication.

2.3.2.2.2 Needed Flight Particle Data

An examination of the data of Section 2.3.2.1 indicates that the flight particle environment has been only loosely defined. A number of additional items would allow a better understanding of the flight particle environments a Spacelab user might expect. This would include a definition of the particle levels during launch and an indication of the relative deposits to be expected on surfaces at various locations and orientations, with correlations to the initial particle population present in the

PLB at launch. For the orbital phase, more detailed correlation of particle generation rates and locations with Orbiter activities would allow users to predict particle generation events and schedule around them. Beneficial descent data would be very similar to the data needed for ascent.

It must be acknowledged that the above information may be difficult and costly to obtain. There are no plans currently to obtain all the above data items. The following section describes the current plans related to STS flight particle environment monitoring.

2.3.2.2.3 Planned Flight Particle Data

Several instruments are currently planned for STS missions which may provide additional flight particle data.

The Interim Operational Contamination Monitor (IOCM) is an instrument package developed by JPL for use on DOD STS missions. Unlike the IECM, which was intended for use to verify STS and Spacelab environments only during early missions, the IOCM is intended to act as an operational contamination monitor during several contamination sensitive DOD missions. The IOCM instruments include 3 TQCMs on the instrument package and 2 TQCMs in other PLB locations, 1 "sticky" QCM for particle collection, a passive particle collector, 2 calorimeters, 2 photodetectors, 2 ion gages, and a materials sample array with 100 samples. It is not yet clear how readily available IOCM data will be, due to its use on DOD STS missions.

For orbital particles, infrared telescopes tend to be the instruments most sensitive to particle contamination. Data on particle contamination can therefore be extracted from performance degradation of these instruments, should they occur. Among the infrared instruments expected to fly on Shuttle is a reflight of the AFGL CIRRIS telescope.

Another instrument, the Martin Ascent Particle Monitor (MAPM), designed specifically for particle contamination detection is being built by Martin Marietta under contract to the Air Force. NASA JSC is providing integration assistance. The device will collect particles during the ascent mission phase. The instrument opens at launch, collects particles on an exposed surface in the Y-Z plane, and then closes at a predetermined PLB pressure level to protect the collected sample. It will be flown with the Evaluation Of Oxygen Interaction with Materials (EOIM) III experiment (see Section 3.1.3.3).

2.3.2.3 Flight Particle Analytical Tools

Analytical tools for use in predicting flight particle environments are currently quite limited. They consist of launch redistribution, orbital trajectory and optical degradation models.

2.3.2.3.1 Launch Particle (Hamberg) Redistribution Model

The commonly used launch particle deposition model was developed by Hamberg of the Aerospace Corporation. It is a simple "smearing" model in which the total Orbiter payload and PLB surface particle population is estimated at launch, and then redistributed in equal densities over all payload and PLB surfaces. In mathematical terms,

$$\frac{N}{A} = \frac{(N_1/A_1)A_1 + (N_2/A_2)A_2 + \ldots (N_n/A_n)A_n}{A_1 + A_2 + \ldots A_n} \quad 2.3\text{-}2$$

where

N/A = final uniform particle density on all surfaces;

N_n = initial number of particles on surface n; and

A_n = area of surface n.

In general, the N_n or the N_n/A_n density is estimated based on the facility fallout data of Section 2.2.2 and the payload ground flow durations. Section 2.2.4 described how such surface densities may be converted to obscuration ratios (OR). Similarly converting Equation 2.3-1,

$$OR = \frac{OR_1 A_1 + OR_2 A_2 + \ldots OR_n A_n}{A_1 + A_2 + \ldots A_n} \quad 2.3\text{-}3$$

where

A_n = area of surface n;

OR = final uniform obscuration ratio for PLB surfaces; and

OR_n = obscuration ratio at launch of surface n.

In general, this approach is expected to yield conservatively high results for given payload surfaces. First, not all particles are expected to be excited or detached by the launch environment. Reference E-14 indicates that smaller particles are not as readily excited by vibroacoustic environments as are large particles. This implies that the total population of particles available for redistribution during launch is not as great as the model assumes. Conservatism may also result from conserving mass within the PLB. Some particles may actually be carried out of the PLB area or concentrated on the PLB vent filters, and thus be unavailable for redistribution. One possible lack of conservatism, however, is the uniform distribution of particles over all PLB surfaces. The acceleration and gravitational forces during launch would be expected to accelerate particles toward the rear of the PLB. Other forces on particles might result in preferred surface locations or orientations for particle deposition. Thus the Hamberg redistribution model may generate conservative average deposits for payload surfaces, but may underestimate the redistribution for specific surfaces in high contamination locations. Data is unavailable to identify and evaluate these areas.

Example 2.3.2.3.1:	Calculation the obscuration ratio on an instrument surface for the SL long module configuration using the Hamberg redistribution model. Assume 1/2 the PLB/SL surface area has an initial OR of 0.015, 1/2 has an initial ratio of 0.0075, and 5000 cm^2 of instrument area has an OR of 0.002.
Solution:	From Table 2.2-18, total PLB and SL area is 4.31 x 10^6 cm^2. From Equation 2.3-2, $$OR = \frac{(.015 + .0075)(2.16 \times 10^6 \text{ cm}^2) + (10^3 \text{ cm}^2)}{(2.16 \times 10^6 + 2.16 \times 10^6 + 10^3) \text{ cm}^2}$$ $$= 0.011$$

2.3.2.3.2 Orbital Particle Tools/Models

Orbital particle models include models to predict the trajectory of particles released by spacecraft while in orbit, and models to predict the effects of particles on the optical properties of a surface or the FOV of an instrument.

The Newtonian Orbital Mechanics and Drag (NOMAD) computer model was developed by Martin Marietta for use in predicting particle trajectories near Skylab (Ref. E-15). The program solved the 2-dimensional equations of motion for a particle in orbit, including drag effects, and compared the motion to that of the source spacecraft. The NOMAD model, now called KORBIT, has recently been upgraded by Martin Marietta to allow solution of particle motion in three dimensions. A similar program has been developed by Lockheed Missiles and Space Company (LMSC) (Ref. E-16). The results of these two models have been compared (Ref. E-17), and are in agreement. The Martin Marietta KORBIT model has recently been used to simulate the LLLTV images of particles from STS-3 described in Section 2.3.2.1.2, and to evaluate the likelihood of particles crossing the field-of-view of the AFGL CIRRIS infrared telescope on a planned STS mission (Ref. E-17).

Particle effects models exist for evaluation of obscuration effects, where particle optical properties are substituted for substrate properties for the area obscured; scattering effects, such as changes in BRDF and hemispherical reflectivity, or background brightness due to cloud reflections; and emission effects where deposited or cloud particles increase the background brightness an instrument observes in the infrared.

Evaluation of obscuration effects involve the most straightforward modeling of these particle effects types. The results from Reference E-9 shown in Section 2.3.2.1.4 showed linear relationships between hemispherical reflectance and solar absorptivity and OR. In general, the degraded absorptivity of a surface may be expressed by:

$$\alpha_{sf} = \alpha_{so} + OR(\alpha_{sp} - \alpha_{so}), \qquad 2.3\text{-}4$$

where

α_{sf} = degraded surface absorptivity;

α_{so} = initial surface absorptivity;

α_{sp} = particle absorptivity; and

OR = obscuration ratio.

Similarly for hemispherical reflectance:

$$\rho_f = \rho_0 + OR(\rho_p - \rho_0), \qquad 2.3\text{-}5$$

where subscripts have the same meanings as in Equation 2.3-4.

For transmitting materials, the obscuration effects (as opposed to scattering effects) may be considered a lessening of the effective light gathering area of a surface. The final effective area is given by:

$$A_f = A_0(1 - OR), \qquad 2.3\text{-}6$$

where

A_f = final effective area;

A_0 = area of clean surface;

and OR is as above. This approach assumes that particles are opaque.

Equations 2.3-4 and 2.3-5 require knowledge of the optical properties of particle contaminants. These properties may be expected to vary greatly, however, depending on the particle source. The data of Reference E-9 shown in Section 2.3.2.1.4 may be used to determine the optical properties of two specific types of controlled particles. These are probably not representative of actual contaminants. An approach likely to yield conservative results for reflectance and α_s degradations is to use the properties of fly ash. Figure 2.3-6 showed solar absorptance changes for fly ash and aluminum oxide contaminants. The sample used for this data was a vapor deposited silver coated quartz second surface mirror with a solar absorptance of < 0.06. Similarly, Figure 2.3-8 shows hemispherical reflectance as a function of incident wavelength for these same materials (complete surface coverage).

Example 2.3.2.3.2:	Calculate the final reflectance and solar absorptivity for a white surface (α_{so} = 0.1, ρ_{so} = 0.9 at 1.2 μm) contaminated to the extent calculated in Example 2.3.2.3.1 (OR = 0.011). Assume the contaminant particles have the properties of fly ash.
Solution:	For solar absorptivity, applying Equation 2.3-4, $$\alpha_{sf} = 0.1 + (0.011)(0.9 - 0.1)$$ $$= 0.109$$ $$(\Delta\alpha = 0.009).$$ For hemispherical reflectance, $$\rho_f = 0.9 + (0.011)(0.16 - 0.9)$$ $$= 0.89$$ $$(\Delta\rho = -0.01).$$

In the case of particles distributed in the volume surrounding a spacecraft, a computer model named CLOUD was developed by Martin Marietta for use on the Skylab program. The CLOUD model calculates particle column densities; Mie scattering parameters for light from the sun, earth and earth albedo; signal attenuation by the cloud; and blackbody emission characteristics. Particle source characteristics must by input into the model. For Skylab, source characteristics were determined by both test and analysis (Ref. E-18).

For the scattering effects of deposited particles, optical system analysis tools such as General Unwanted Energy Rejection Analysis Program (GUERAP) II (Ref. E-19) and Arizona's Paraxial Analysis of Radiation/Program for the Analysis of Diffracted Energy (APART/PADE) (Ref. E-20) may be used to analyze the impact of particles on a system's performance, if the degradation of the individual component surfaces can be determined. Changes in BRDF due to particles may also be extended to other properties, such as diffuse reflectivity, if the BRDF changes can be determined or estimated.

2.3.2.4 Flight Particle Protection Techniques

If it is determined that an experiment is sensitive to the particle contamination levels expected during ground, launch, orbital or descent operations, protection techniques may be appropriate. Techniques include:

1) Ground covers, removed shortly before launch;

2) Special cleaning and inspection procedures for the payload and PLB (KSC sensitive and highly sensitive classifications);

3) Selection of PLB location and orientation relative to likely launch particle sources;

4) Active covers (openable, or openable and closable);

5) Experiment scheduling around contamination events
   - late in mission
   - unilluminated times
   - engine/vent suppression;

6) Design tolerance to allow degradation; and

7) Constraints on pointing to avoid likely regions for particles.

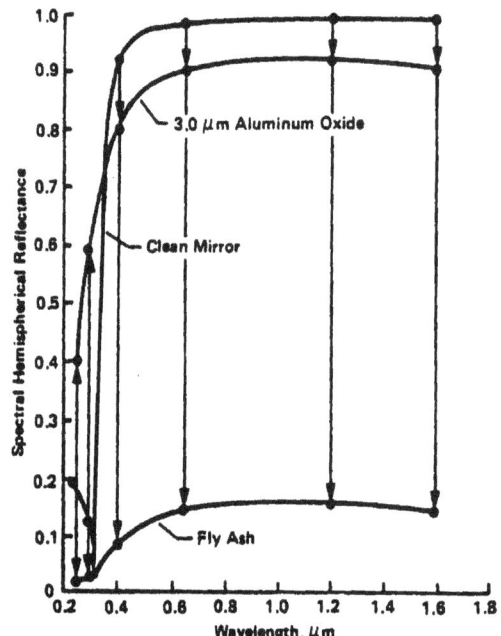

Figure 2.3-8 Mirror and Contamination Reflectances (Ref E-9)

The first two protection techniques for flight particle protection are actually ground activities. They are included here, however, since the ground activities establish part of the particle reservoir for flight, and the effects of ground particles are usually most critical during orbital phases. Covering the critical surface itself can minimize the background level at the time of launch. As described in Section 2.3.2.3.1, however, a large population of particles in the bay can degrade a clean surface during launch. The second technique serves to decrease the total PLB particle population, and thus is more effective at controlling launch particles.

The third technique is currently not experimentally justified, since the launch data is not detailed enough to show location and orientation dependencies. In theory, however, launch particle deposition should be lowest in the forward part of the PLB and on surfaces vertical during launch.

Active covers give experimenters the greatest control over their surface cleanliness. The use of such a device may greatly complicate hardware and operation design, however, particularly if crew or ground activation is required. The mechanism also provides another possible failure point. This was the case with the CIRRIS infrared telescope of STS-4, when the cover did not deploy, and no data was obtained.

Experiment scheduling is probably the most effective technique for dealing with orbital particles crossing fields-of-view. Some experiments, however, may not have the flexibility to select operation times. Also, the Camera-Photometer data of SL-1 does not show a steady decrease in particles with mission time. It is possible that this will also be the case in other Spacelab missions, with a large number of experiments located in the PLB. The length of time that engines and vents may be inhibited is limited, and this also interferes with a scheduling approach to particle protection.

The final approach listed for protection from orbital particles is to include greater tolerances in the design of hardware elements. For example, if degradation of a thermal control surface due to particles is anticipated the surface could be sized to allow it to meet its requirements even in a degraded state. In addition, instruments can be designed to discriminate between particles and other sources, as was done with the SL-2 IRT.

2.3.2.5 <u>Key Flight Particle Technical Personnel</u>

	Organization	Phone
Edgar R. Miller (IECM)	MSFC	(205) 544-7752
R. C. Linton (OEM/PSA)	MSFC	(205) 544-2526

	Organization	Phone
K. S. Clifton (Camera-Photometer)	MSFC	(205) 544-7725
Lubert J. Leger	JSC	(713) 483-2059
Steve Jacobs	JSC	(713) 483-3561
Jack Triolo	GSFC	(301) 344-8651
Fred Witteborn (SIRTF/IRT)	ARC	(415) 694-5520
Jack Barengoltz (Shuttle Env. Particle Subcommittee Chair)	JPL	(818) 354-2516
Carl R. Maag (IOCM)	JPL	(818) 354-6453
Lyle E. Bareiss	Martin Marietta	(303) 977-8713
Frank J. Jarossy (Particle Effects)	Martin Marietta	(303) 977-8716
Milton A. Hetrick (MAPM)	Martin Marietta	(303) 977-1907
John H. Raab (cleanliness/redistribution)	Martin Marietta	(303) 977-1878
Kathleen Muscari (KORBIT)	Martin Marietta	(303) 977-8672
Otto Hamberg (Redistribution/Effects)	Aerospace Corp.	(213) 648-5821
Gene N. Borson	Aerospace Corp.	(213) 648-6943
Jerry L. Weinberg (SIA Experiment)	Univ. of Fl.	(904) 392-5450

2.3.2.6 Flight Particle References

E-1. Miller, E. R., ed., "Induced Environment Contamination Monitor Preliminary Results from the Spacelab 1 Flight", NASA TM-86461, MSFC, August 1984.

E-2. Miller, E. R., "STS-2,-3,-4 Induced Environment Contamination Monitor (IECM)", NASA TM-82524, MSFC, February 1983.

E-2A. Miller, E. R., "Update of Induced Environment Contamination Monitor Results", AIAA-83-2582-CP, October 31, 1983.

E-3. Miller, E. R., "STS-2 Induced Environment Contamination Monitor (IECM) - Quick Look Report", NASA TM-82457, MSFC, January 1982.

E-4. Linton, R. C., E. R. Miller and M. Susko, "Passive Optical Sample Assembly (POSA): Final Report", NASA TM-82466, MSFC, August 1981.

E-5. Bareiss, L. E., "Workshop on STS Payloads Environmental Data: Contamination Panel Summary", Martin Marietta, 9 June 1983.

E-6. Dixon, D. G., W. C. Carey and J. A. M. McDonnell, "Contamination By Fibers on Space Shuttle Flight OSS-1 Microabrasion Foil Experiment", Journal of Spacecraft, Vol. 21, No. 4, April 1984, University of Kent, published by AIAA.

E-7. Miller, E. R. and R. Decher, "An Induced Environment Contamination Monitor for the Space Shuttle", NASA TM-78193, MSFC, August 1978.

E-8. Barengoltz, J., F. Kuydendall, and C. Maag, "The Particle Environment of STS-3 as Observed by the Cargo Bay Television System", JPL, 25 October 1982.

E-9. Hamberg, O. and F. D. Tomlinson, "Sensitivity of Thermal Surface Solar Absorptance to Particulate Contamination", AIAA-71-473, Aerospace Corporation, published by AIAA, April 1971.

E-10. Young, R. P., "Degradation of Low Scatter Mirrors by Particle Contamination", AEDC-TR-74-109, ARO Inc., January 1975.

E-11. Leinert, C., "Stray Light Suppression in Optical Space Experiments", Applied Optics, Vol. 13, No. 3, March 1974.

E-12. "Mirror BRDF Plot", Hughes Aircraft Co., Informal Communication, no date.

E-13. "The Contamination Environment of STS Mission 51-C as Measured by the Interim Operational Contamination Monitor (IOCM)" (U), Jet Propulsion Laboratory, August 1985.

E-14. Hamberg, O., "Prelaunch and Orbiter Bay Contamination Control at KSC", Aerospace Memo 78-5124.17-15, 16 November 1978.

E-15. Sherrard, M. L. and L. E. Bareiss, "Apparent Trajectories of Contamination Particles in the Spacecraft Environment", NASA New Technology Submission MFS-22844, Martin Marietta, 13 July, 1973.

E-16. Lee, A. "Particle Dispersion Around a Spacecraft", AIAA-83-0243, LSOC, 1983.

E-17. Muscari, K. "The Particle Environment Around Sensitive Payloads", AIAA-85-0955, Martin Marietta, 19 June 1985.

E-18. Sherrard, M. L., F. J. Garlitz, and F. J. Jarossy, "Cloud Effects Math Model Report", ED-2002-1372, Rev. A, Martin Marietta, 30 September 1972.

E-19. Likeness, B. K., "GUERAP III: General Unwanted Energy Rejection Analysis Program, User's Manual", Honeywell, Inc., July 1978.

E-20. Breault, R. P., "User's Manual for PADE-APART Version 6B", Breault Research Organization, February 1980.

2.3.3 Flight Molecular Environment

2.3.3.1 Available Data

In this section an overview of published flight molecular contamination data will be presented. For further details the cited references should be consulted. The presentation will generally follow the order of the IECM instrumentation descriptions in Section 2.3.1.2.

2.3.3.1.1 Ascent Phase

The Air Sampler (IECM03) was used to "grab" the gaseous environment during ascent and descent.

After collection, the samples were analyzed in ground laboratories by gas chromatography/mass spectroscopy (GC/MS). Figure 2.3-9 shows the collection timeline on STS-2 during ascent (Ref. F-1). Table 2.3-3 lists some of the compounds and the quantities measured; a listing of 127 entities was generated some of which could not be identified. The ascent A and B terminology refers to the two samples collected during the STS-2 ascent. A summary of ascent Air Sampler results from the STS-2, -3, and -4 missions is presented in Table 2.3-4 (Ref. F-2). Detection method B in Table 2.3-4 consisted of Electron Spectroscopy for Chemical Analysis (ESCA) used for reactive species that react with the platelets (see Sec. 2.3.1.2). The Air Sampler results obtained during the STS-9/Spacelab 1 (STS-9/SL-1) mission are presented in Table 2.3-5 (Ref. F-3).

The TQCM (IECM07) measured condensables in the payload bay, but during ascent the temperatures of the sensors are not controlled. Figure 2.3-10 shows a comparison of the mass accumulation for the +Y (right) axis sensors during ascent of flights STS-2, -3, and -4 (Ref. F-4). It can be seen that mass accumulates during ascent, reaches a peak in less than 1 minute, and as orbital altitudes are reached the accumulated mass begins to desorb and in most cases reaches a level which is below the original level. Table 2.3-6 summarizes the ascent data for the five sensors for the three flights (Ref. F-4).

Table 2.3-3
IECM—Air Sampler Organics Detected in Most Significant Quantities* on STS-2 (Ref F-1)

Compound	Ascent A, μ gm	Ascent B, μ gm
$C_9H_{16}O$ (2-Isononenal)	—	0.22
4-Methyl-1-Pentene	1.5	3.2
1, 1, 2-Trichloro-1, 2, 2-Trifloroethane	—	—
Methyl Benzene	4.5	0.86
Dibromochloromethane	—	—
Hexamethylcyclotrisiloxane or Similar Compound	7.2	0.88
A Carboxylic Acid Ester, No Satisfactory Match	7.5	11.0
Benzenedicarboxylic Acid, Di C_{14} Ester	2.2	3.4
Nonadecane	1.2	1.5
1, 2-Benzenedicarboxylic Acid, Diethyl Ester	10.0	15.0
Aromatic Hydrocarbon (Possibly Oxygenated)	3.8	6.8
	In Approx 110 Std CC	In Approx 110 Std CC

*Subsequent control analyses show that these quantities must be lowered significantly.

Table 2.3-4
Summary of Results of Air Sampler Contaminants during Ascent from STS-2, -3, and -4 (Ref F-2)

Mission Phase	Species	Detection* Method	Observed
Ascent	Volatile Hydrocarbons†	A	50 ppm by Weight 10 ppm by Volume
Ascent	Reaction HC1	B	None Detected to ppm Sensitivity

*A — Concentration on absorbent; postflight GC/MS analysis.
 B — Reaction with silver oxide/hydroxide surfaces.
 C — Reaction with ruthenium trichloride surfaces.

†Covers C_9 to C_{24} range and uses ~ C_{12} as average molecular weight to obtain ppm by volume.

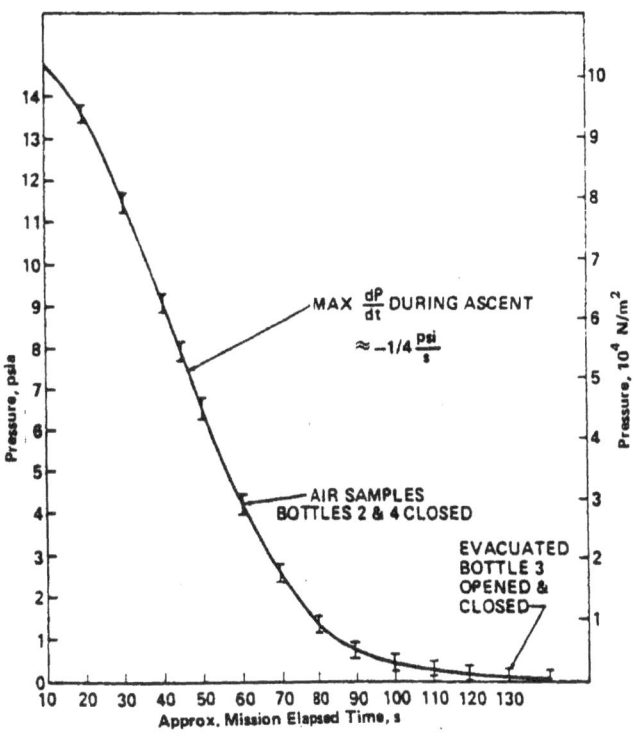

Figure 2.3-9 Pressure During Ascent of STS-2 (Ref F-1)

Table 2.3-5
Species from Ascent Air Sampler on STS-9/Spacelab 1
(Ref F-3)

Peak Number	Area Percent	Amount in 10^{-9} g per Substance	Substance Identity
1	3.8	45	Hexamethylcylotrisiloxane
2	1.2	14	Cylic Hydrocarbon
3	4.1	49	Unidentified
4	2.1	25	Low MW Alcohol or Ketone
5	8.4	100	C_7 Branched Alkane
6	2.8	34	C_7 Branched Alkane
7	0.2	2	Unidentified
8	1.1	13	C_8 Branched Alkane
9	1.0	12	C_9 Branched Alkane
10	2.1	25	Octamethylcylotetrasiloxane
11	2.5	30	Branched Alkane
12	4.9	59	Aromatic Ketone (1-Phenyl Ethanone?)
13	3.0	36	Branched Alkane (C_9?)
14	2.9	35	Branched Alkane (C_{10}?)
15	2.8	33	Branched Alkane
16	1.3	16	Branched Alkane
17	2.6	31	(Alcohol or Diol?)
18	1.6	19	Branched Alkane
19	3.6	43	Naphthalene
20	1.0	12	(Ester?)
21	4.0	48	Unidentified
22	1.0	12	Tridecane
23	1.5	18	Tetradecane
24	0.9	11	Ester or Diester
25	1.7	20	Unidentified
26	3.0	36	Branched Alkane
27	2.3	28	Substituted Aromatic Comp
28	1.0	12	Unidentified
29	2.0	24	Unidentified
30	5.6	67	Branched Alkane
31	1.4	18	Unidentified
32	2.8	34	Branched Alkane
33	4.7	56	Branched Alkane
34	3.5	42	Branched Alkane
35	3.7	44	Branched Alkane
36	4.7	56	Branched Alkane
37	2.5	32	Branched Alkane
38	1.0	12	Branched Alkane

The results from the cryogenic quartz crystal microbalances, CQCM (IECM08), for the ascent phases of the three flights are presented in Table 2.3-7 (Ref. 6). In most cases at the end of the measurement it appears that there is a negative mass accumulation.

Table 2.3-7 CQCM Net Molecular Mass Accumulation Rates during Ascent Phase (Ref F-4)

	MET, h min	Δ Time, min	Sensor -Z1			Sensor -Z2		
			Sensor Temp, °C	Mass Change, ng cm^{-2}	Mass Accum Rate, ng cm^{-2}/h	Sensor Temp, °C	Mass Change, ng cm^{-2}	Mass Accum Rate, ng cm^{-2}/h
STS-2	000 00 / 000 37	37	23 / 21	67	109	23 / 20	-6	-10
STS-3	000 00 / 000 37	37	23 / 20	-4	-8	24 / 21	-28	-47
STS-4	000 00 / 000 17	17				25 / 22	17	61

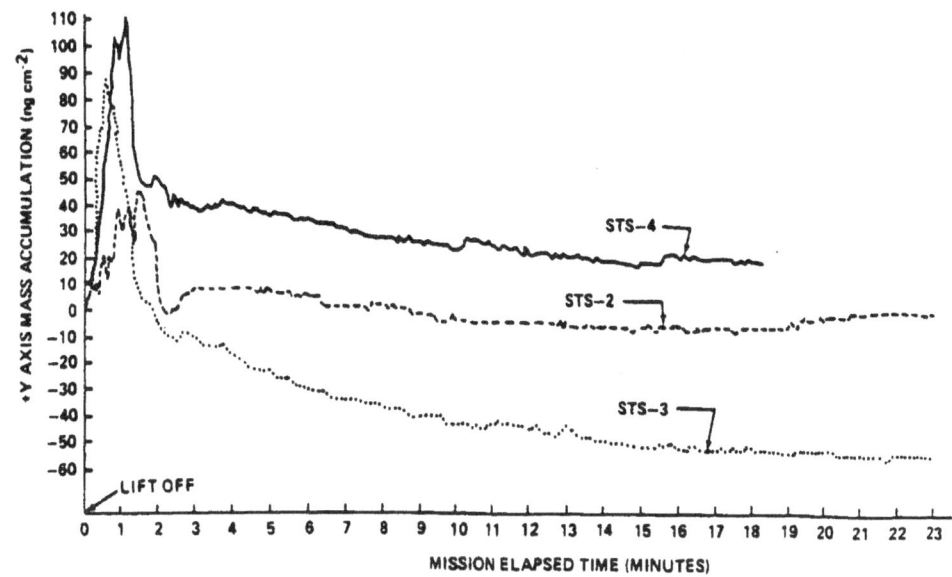

Figure 2.3-10 Comparison of +Y Axis Mass Accumulation during Ascent (Ref F-4)

Table 2.3-6
Mass Accumulation by TQCM during Ascent
Phase (Values Shown in ng cm^{-2}) (Ref F-4)

STS-2—Ascent

Sensor Axis	Max	Time	Mission Elapsed Time, Min				
			2	10	20	30	37
+Y	+45	1.5	-2	-6	-2	-43	-149
+X	109	0.8	-20	-16	-62	-69	-179
-Z	33	0.8	-27	+22	+61	+81	+94
-X	134	1.0	-8	-20	-8	-11	-11
-Y	129	2.0	—	+6	-739	-465	-674

STS-3—Ascent

Sensor Axis	Max	Time	Mission Elapsed Time, Min				
			2	10	20	30	37
+Y	+87	0.7	-2	-45	-58	-69	-72
+X	+109	0.6	-16	-47	-67	-76	-81
-Z	+111	0.6	-9	-16	-20	-25	-25
-X	+145	0.8	-9	-61	+75	+70	+75
-Y	+145	0.8	+8	0	+2	-12	-27

STS-4—Ascent

Sensor Axis	Max	Time	Mission Elapsed Time, Min				
			2	5	10	15	18
+Y	112	1.2	+47	+36	+22	+17	+17
+X	195	0.9	-114	-265	-535	-449	-412
-Z	41	0.8	-2	-11	-25	-45	-53
-X	144	1.0	+23	+6	-3	-9	-12
-Y	192	0.9	+11	+20	+43	+28	+27

2.3.3.1.2 On-Orbit Phase

From analysis of results from the Passive Sample Array (IECM05) and the Optical Effects Module (IECM06) on STS-2, -3, and -4, there appeared to be no significant evidence of molecular film deposition (Ref. F-5). Most of the measured degradation was probably due to the effects of adhering particles which was attributed to the post-landing ferry-flight environment. Similar results were obtained on STS-9/SL-1 (Ref. F-6).

During missions STS-2, -3, and -4, the TQCM (IECM07) sensor accumulated mass periodically at four preprogrammed temperature settings: $+30°$, $0°$, $-30°$ and $-60°C$. Between each measurement period the sensors underwent a $+80°C$ cleanup mode. On STS-2 and -3 most accumulation rates were between 0 and 50 ng/cm^2 hr with some negative values being recorded at each temperature. On STS-4 the majority of measured deposits were less than 25 ng/cm^2 hr and many negative values were recorded (Ref. F-4). The contamination measured on STS-9/SL-1 was significantly greater than on the three early missions. Figure 2.3-11 shows the TQCM measurement for all five sensors on the STS-9 mission (Ref. F-7). The +X (fore) sensor clearly accumulated the most mass during any period of the mission.

Rapid and large changes in mass accumulation can be related to particular events on a mission. Figure 2.3-12 shows the mass accumulation recorded during a Reaction Control System (RCS) engine firing on STS-3 (Ref. F-4). Diurnal variations in mass accumulation have also been observed and Figure 2.3-13 shows such effects on mission STS-9/SL-1 (Ref. F-7). (Similar variations were also observed on STS-3.) The largest seem to occur with the +Y sensor.

On STS-3 the TQCM of the CMP, contamination monitor package (Section 2.3.1.2.2) recorded accumulations in a predictable pattern based on the temperature profile of the payload and its surroundings (Ref. F-8). There was a strong dependence on the temperature of the bay which, in turn, depends on the Orbiter attitude towards the sun. In the tail-to-sun position (TTS), accretion rates on $0°C$ TQCM surfaces began as high as 100 ng/cm^2 hr during the first hour and fell rapidly to a few ng/cm^2 hr (both positive and negative) almost immediately. Low rates existed even when the TQCMs were set to $-30°C$ as the cargo bay continued cooling. In the nose-to-sun attitude (NTS), which is a slightly warmer condition, the same low rates at 20 and $-30°C$ were seen. In the bay-to-sun attitude (BTS), accretion rates began rising to nearly 100 ng/cm^2 hr for the -X (aft) sensor at $0°C$. The other sensors showed rates in the vicinity of 20 ng/cm^2 hr. At $20°C$, all of the TQCMs showed accretions in the vicinity of 20 ng/cm^2 hr. Table 2.3-8 shows the accumulations for a number of periods during the STS-3 mission. TQCM 3 was the -Z sensor, while TQCM 4 pointed in the +Y direction.

The CQCM (IECM08) mass accumulation results for STS-2, -3, and -4 are summarized in Table 2.3-9 (Ref. F-4). The results for STS-9/SL-1 for the -Z sensor is shown in Figure 2.3-14 from lift-off to 177 hr MET (Ref. F-7). Contamination deposition measured by the CQCM on this axis was less than 4 ng/cm^2 hr.

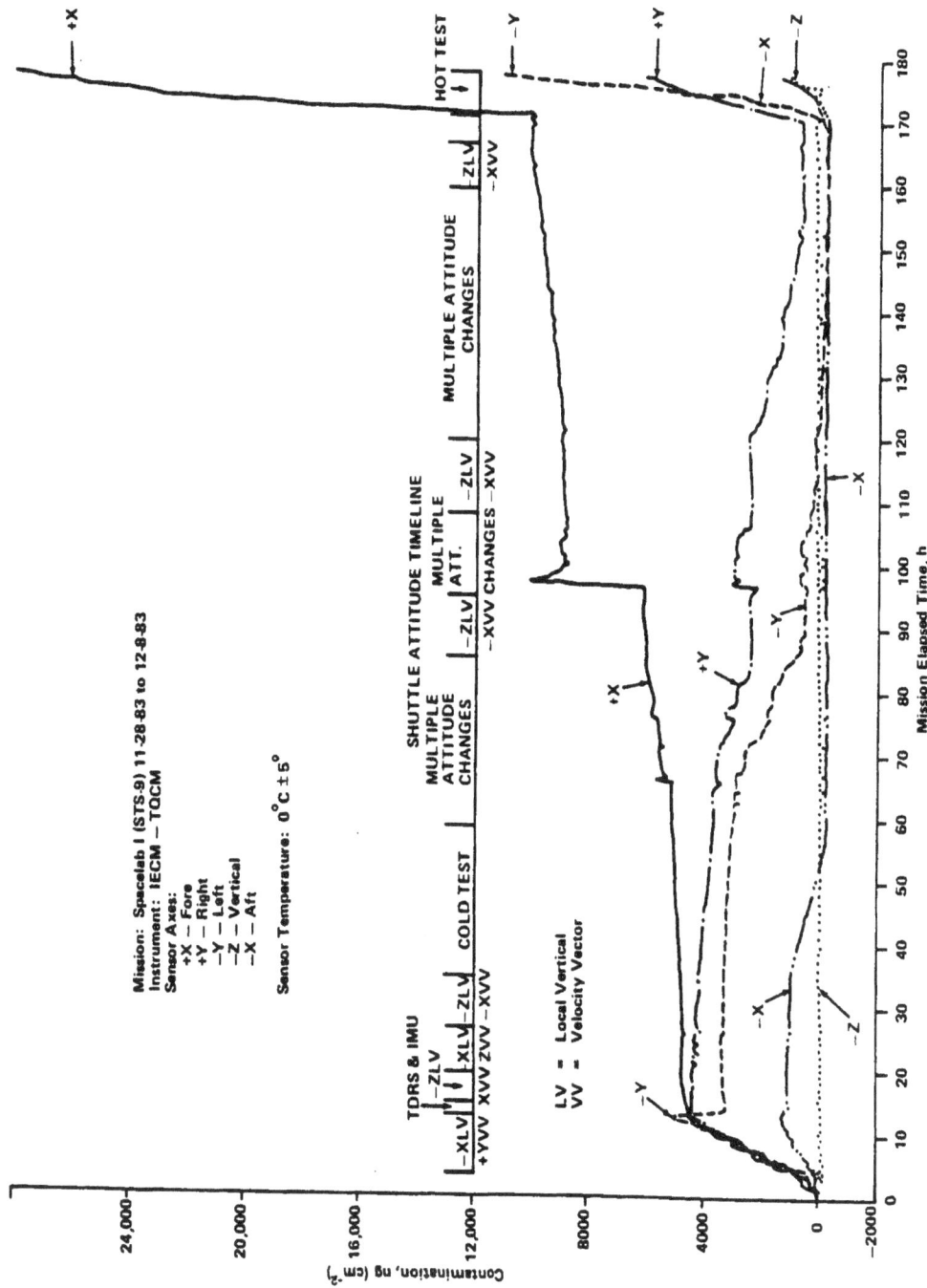

Figure 2.3-11 Surface Contamination as Measured by TQCM on Spacelab 1 (Ref F-7)

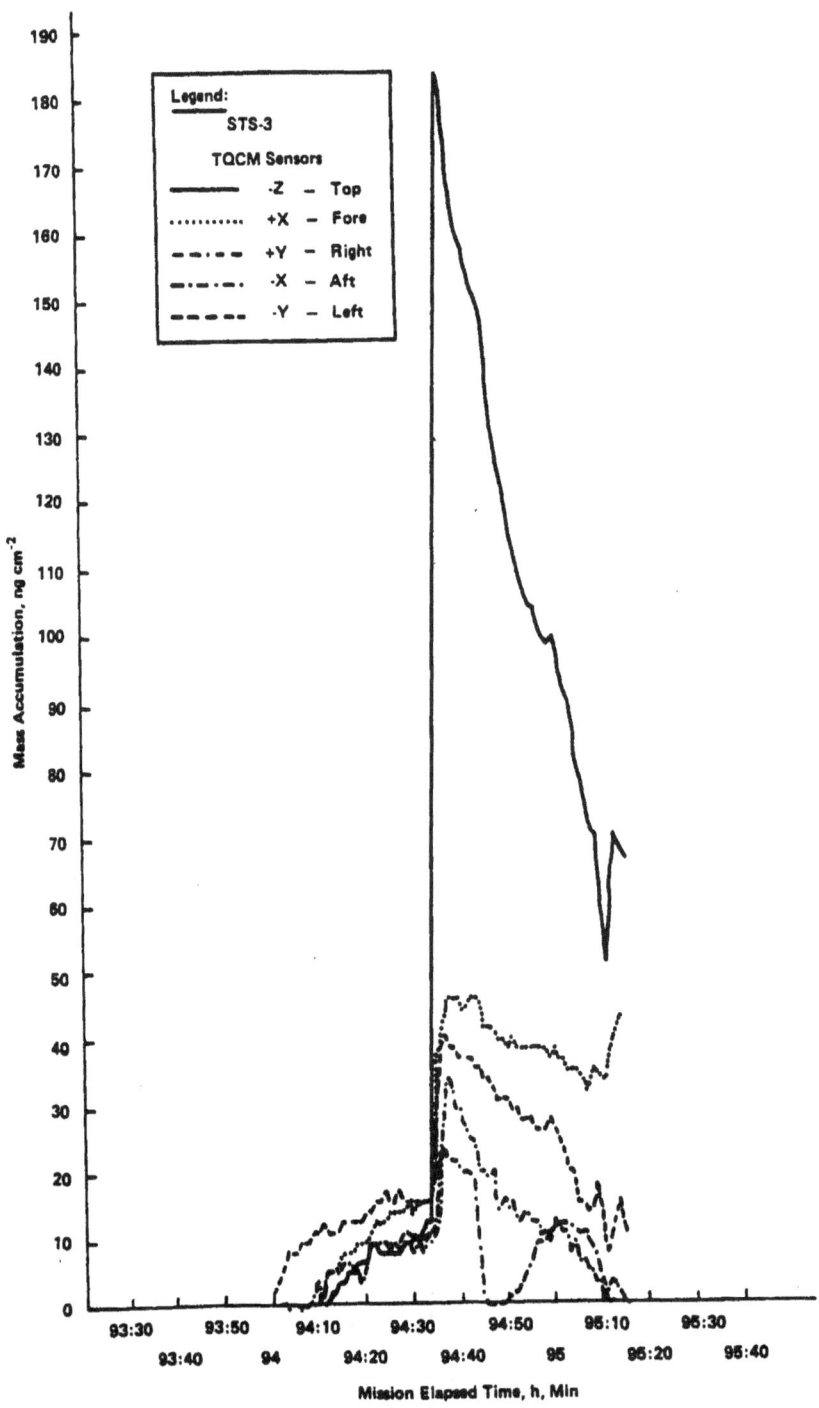

Figure 2.3-12 Mass Accumulation during STS-3 L2U Engine Firing (Ref F-4)

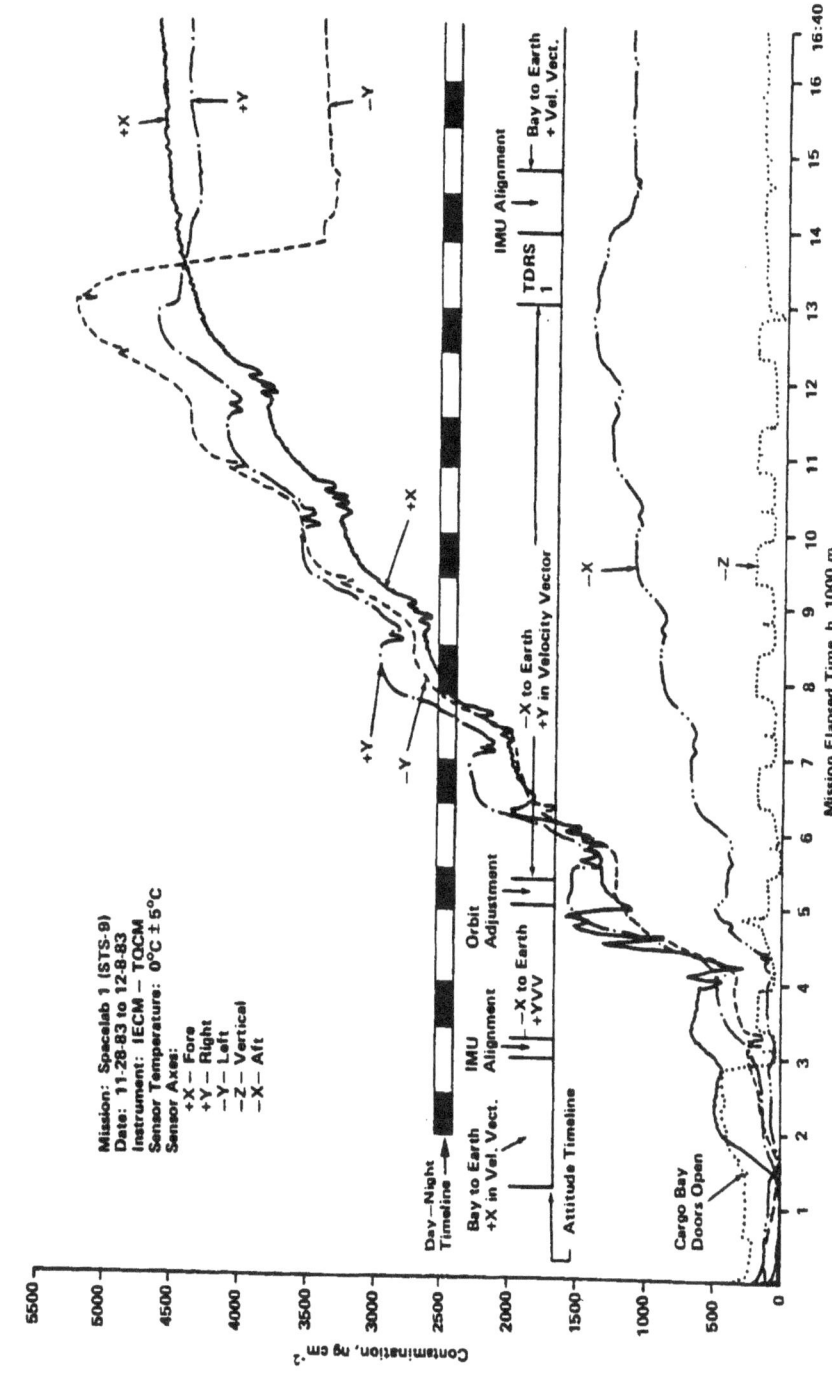

Figure 2.3-13 IECM TQCM Surface Contamination on Spacelab 1, First 16.7 Hours (Ref F-7)

Table 2.3-8 Accretion Rates Indicated by the CMP during the STS-3 Mission (Ref F-8)

MET		Orbiter Attitude	TQCM Temperature, °C	TQCM Accretion Rates, ng cm^{-2} h^{-1}			
From	To			-X	-Y	-Z	+Y
000:12:50:16	000:15:50:25	TTS	+2	+4	+3	-2	+2
000:18:39:25	000:21:39:24	TTS	+1	+3	-1	-1	+3
001:01:00:30	001:04:00:38	TTS	-1	+3	0	-1	+3
001:05:15:19	001:06:45:42	TTS	-30	-2	0	+1	+5
001:06:45:42	001:08:15:27	TTS[a]	-30	-10	-8	+17	-2
001:12:45:22	001:15:45:29	PTC	-29	-9	+3	0	+3
002:06:30:07	002:09:30:18	NTS	+15	+7	-2	-13	+4
002:11:00:02	002:14:00:36	NTS	+14	-1	+4	+2	+2
003:02:00:23	003:03:30:08	NTS	-29	0	-12	-48	-2
003:04:45:38	003:07:45:43	NTS	-29	-2	-1	-9	-1
003:11:00:07	003:14:00:20	NTS	-29	-2	+1	+2	0
003:15:30:08	003:18:30:21	NTS	-29	+4	-1	+2	-2
004:03:00:05	004:06:00:15	NTS[b]	-29	-3	-13	+7	-12
004:13:20:00	004:16:20:08	NTS	-29	+1	0	+6	-2
005:00:05:07	005:03:05:05	NTS	-29	-3	+13	-4	-5
005:10:15:18	005:13:15:28	BTS	+1	+51	-9	+11	+32
005:15:00:17	005:18:00:30	BTS	+1	+81	+14	+2	+18
006:01:15:33	006:04:15:03	BTS	+17	+7	+12	+14	+9
006:05:45:28	006:08:45:22	BTS	+17	+26	+18	+16	+87[d]
006:10:00:05	006:13:00:20	PTC	+1	-2	0	+2	-3

Notes:
The values are based upon the differences between two groups of five points each either one or two orbits apart. The times given are the center points of the five. The temperatures can vary by 2°C from the value given; in the case of TQCM3, instability often caused oscillations of as much as ± 12°C.

[a] Includes payload bay door opening and closing tests.
[b] After a bakeout.
[c] Temperature change took place during this period.
[d] Temperature of TQCM 4 was -27°C.

Table 2.3-9 CQCM Summary: Net Molecular Mass Accumulation Rates (Ref F-4)

	-Z1				-Z2			
	Frequency Change, Hz	Mass Change, ng cm^{-2}	Time Interval, min	Mass Accumulation Rate, ng cm^{-2} h^{-1}	Frequency Change, Hz	Mass Change, ng cm^{-2}	Time Interval, min	Mass Accumulation Rate, ng cm^{-2} h^{-1}
From Liftoff to Power Down								
— STS-2	-109	-170	3,285	-3	-372	-580	3285	-11
— STS-3	119	186	11,580	1	144	225	11,580	1
— STS-4	-157	-245	10,202	-1	-32	-50	10,202	-0.3
From Minimum Frequency to Maximum Frequency in Orbit								
— STS-2	212	331	2,787	7	245	382	1,941	12
— STS-3	406	638	8,858	4	393	613	10,009	4
— STS-4					287	448	2,458	11
From Minimum Frequency to Final Frequency in Orbit								
— STS-2	174	271	2,957	5	147	229	2,952	5
— STS-3	340	530	11,488	3	290	452	11,276	2
— STS-4					183	285	9,705	-1

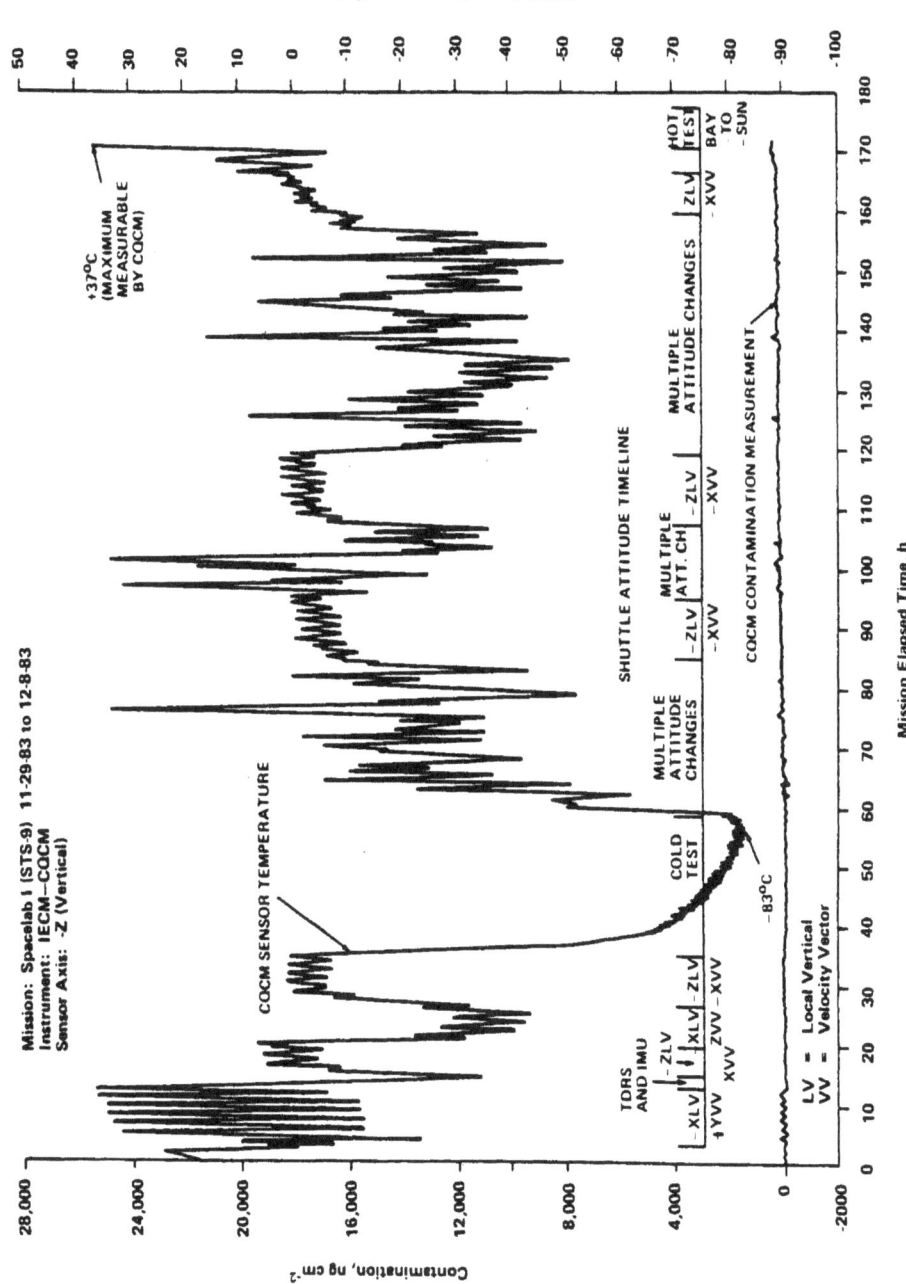

Figure 2.3-14 IECM CQCM Surface Contamination Measurements and Sensor Temperature (Ref F-7)

For on-orbit measurements the quadrapole mass spectrometer (IECM10) was normally oriented to view along the -Z axis so that the measured constituents were mostly a consequence of scattering from the ambient atmosphere. Gaseous atoms and molecules of 1 through 150 amu were sampled for 2 seconds with a full spectrum being obtained in 3 seconds. Particular emphasis was on H_2O contamination so it was sampled on a continuous basis between scans of the entire spectrum.

Both STS-2 and -4 suffered from heavy rainstorms on the launch pad. The early part of the STS-4 mission (<4 hrs MET) showed the highest water return flux values of the three flights. It decreased with a time constant (1/e) of about 10 hours. Table 2.3-10 shows the mass spectrometer H_2O return flux measurements as well as calculated column densities. The table also shows H_2O on the STS-9 flight (Ref. F-2).

Table 2.3-10
H_2O Return Flux by IECM-Mass Spectrometer and Calculated Column Densities (Ref F-2)

Mission	Return Flux ($/cm^2/sr/s$)		Column Density ($/cm^2$)	
	*Maximum	Final	Maximum	Final
†STS-2	1.3×10^{14}	1.8×10^{13}	2.0×10^{13}	2.7×10^{12}
STS-3	9.8×10^{11}	2.6×10^{11}	1.5×10^{11}	4.0×10^{10}
STS-4	2.1×10^{14}	6.6×10^{12}	3.2×10^{13}	1.0×10^{12}
STS-9			2.0×10^{12}†	

*Except for PRCS firings and payload bay door closings.
†The values are considered upper limits.

Excess water from fuel cells is dumped periodically at rates of about 68 kg/hour with a dump usually lasting for about an hour. Of the 25 water dumps during the STS-2 and STS-4 missions, only one, occurring at 118 hours MET, on the STS-2 mission was clearly correlated with the mass spectrometer response. This is shown in Figure 2.3-15 (Ref. F-9). Further, the mass spectrometer did not unambiguously detect any increase in H_2O return flux during Flash Evaporator System (for cooling) operations. (If any water release froze into ice crystals, as might be expected, water would not be detected by the mass spectrometer.)

Water contamination can, however, be correlated with Shuttle surface temperatures (Ref. F-10). Figure 2.3-16 shows the variation of water with MET as well as the AFGL mass spectrometer sensor temperature. This temperature profile was identical to profiles of thermistors placed throughout the pallet in the bay where the latter showed excursions of at least 100 °C at the peaks. Thus, it is probable that the water output directly reflects the temperature induced outgassing or desorption of spacecraft surfaces.

Figure 2.3-17 shows a mass spectral scan at 7.2 hours MET on STS-3 (Ref. F-10). It is not sufficiently well resolved at most mass numbers for a definitive analysis of contaminants. It is, however, remarkably clean above 50 amu. On STS-4, Freon 21 was a significant contaminant, implying a leak in a cooling loop. Helium is common during all three flights, and it is probable that small leaks in the many helium pressurized systems account for much of the helium observed (Ref. F-11). Figure 2.3-18 shows the pressure rise in the payload bay during a door closing on the STS-3 flight (Ref. F-11).

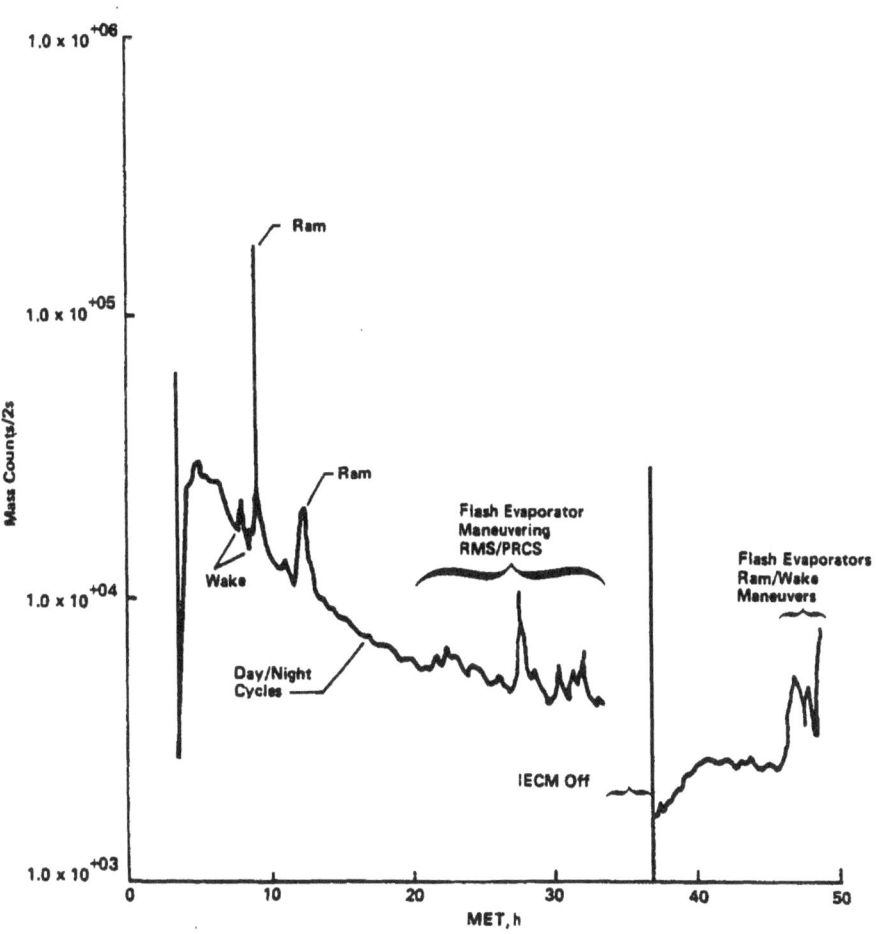

Figure 2.3-15 IECM Mass Spectrometer, Flight STS-2: Mass Counts vs Time at amu 18 (Ref F-9)

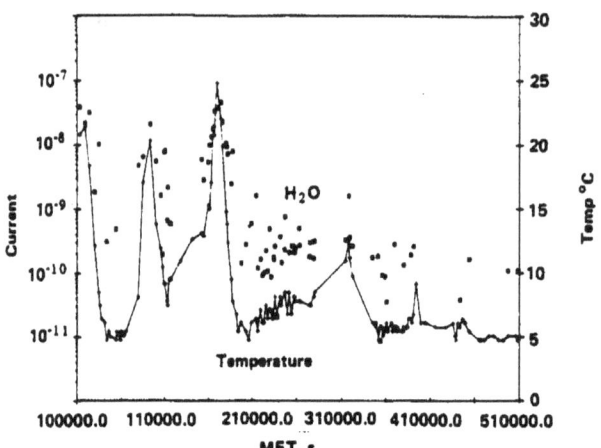

Figure 2.3-16
The Average H_2O Current and Sensor Temperature Measurements Throughout Flight (Ref F-10)

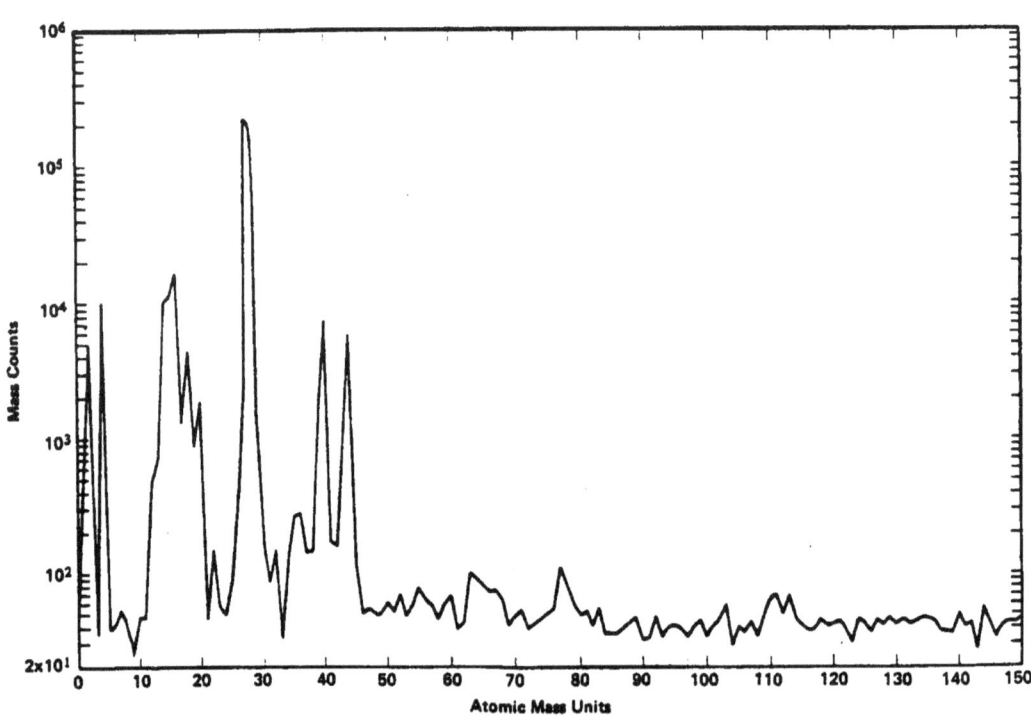

Figure 2.3-17 IECM Mass Spectrum, STS-3, 7.2 h MET (Ref F-11)

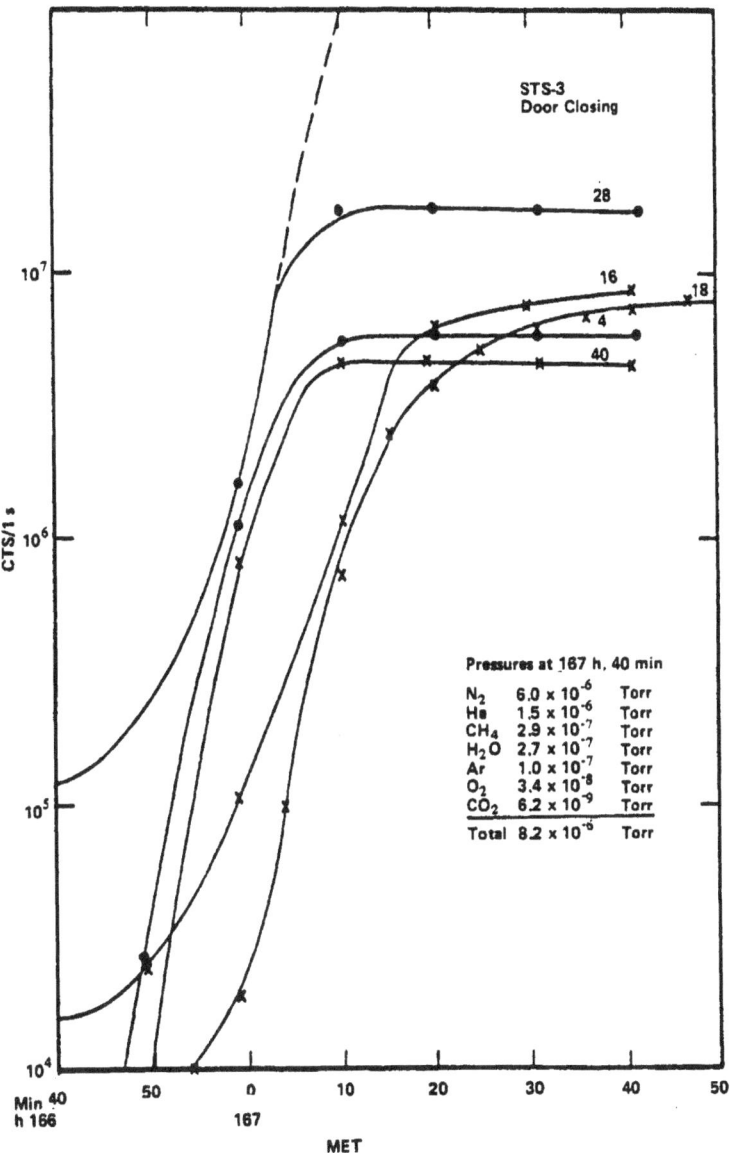

Figure 2.3-18
Pressure Rise and Composition by IECM Mass Spectrometer in the Payload Bay During the Door Closing Exercise at 167 hr MET on STS-3 (Ref G-12)

On STS-4 the IECM was picked up by the RMS and maneuvered to look inward at the payload bay and other surfaces. The survey was taken between 45.5 and 47.8 hours MET in bay-to-sun attitude. A total of 15 different configurations were achieved of which nine are depicted in Figure 2.3-19 (Ref. F-12). The field-of-view of the mass spectrometer, 10° half angle, is shown in positions 11 and 18. In all positions the IECM was located on the center line of the Y axis of the Shuttle. The average value at each position of three of the observed contaminants is shown in Figures 2.3-20, -21, and -22. In Figure 2.3-22, positions 11 and 17 seem to localize the source of the Freon leak in the vicinity of the aft bulkhead and the tail root. The helium survey shows a very large source at position 18 (see top of Figure 2.3-21).

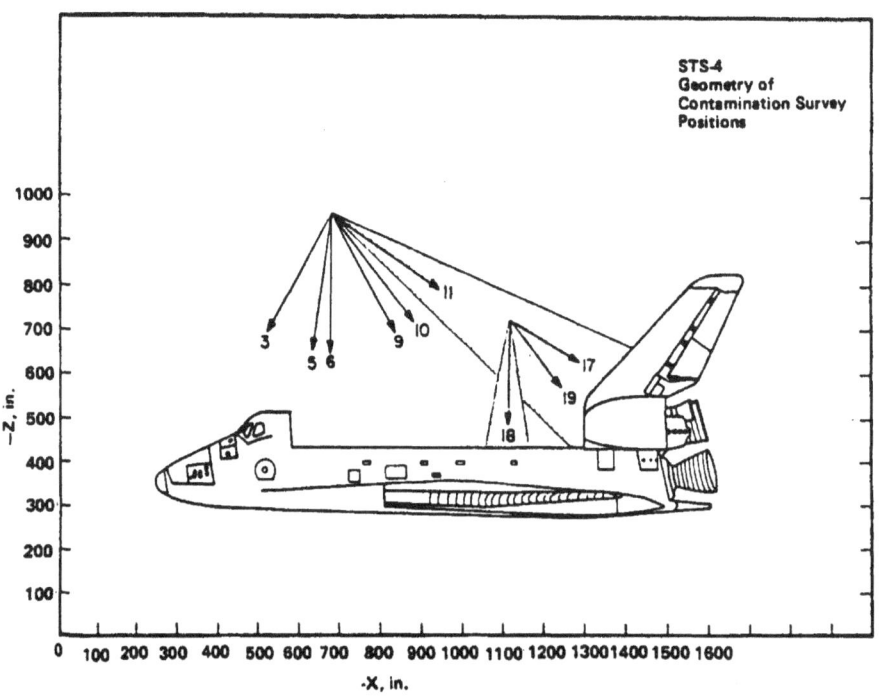

Figure 2.3-19 STS-4 Geometry of Contamination Survey Positions (Ref F-12)

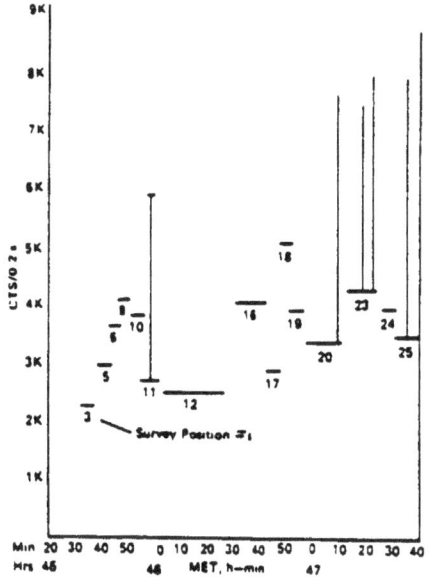

Figure 2.3-20 STS-4 H_2O Counts during Contamination Survey (Ref F-12)

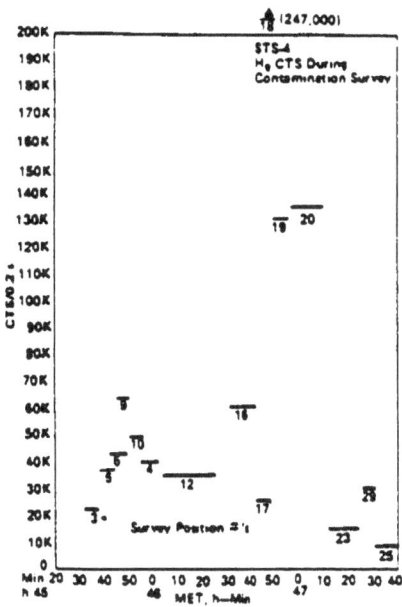

Figure 2.3-21 STS-4 He Counts during Contamination Survey (Ref F-12)

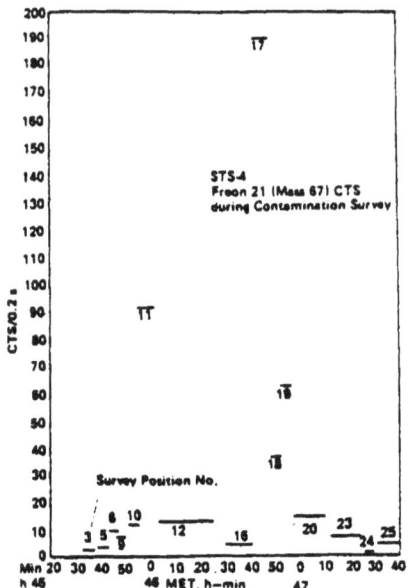

Figure 2.3-22
STS-4 Freon 21 (Mass 67) Counts during Contamination
Survey (Ref F-12)

An unexpected measurement by of the mass spectrometer on STS-9/SL-1 was a large peak for methane during PRCS thruster tests. A thruster firing event on STS-4 viewed directly during the RMS survey, is shown in Figure 2.3-23 (Ref. F-12) methane is not predicted as a contaminant and is ascribed to catalytic production over the zirconium oxide getters of the collimator of the mass spectrometer from unburned monomethyl hydrazine. This assertion requires additional study but does seem to explain the results.

On STS-9 the mass spectrometer results were quite similar to earlier flights. However, during low temperature excursions, depicted in Figure 2.3-24 the instrument became slightly detuned moving a given amu to a lower value. Very little, if any, data was lost as a result of this problem, but the data reduction process become more complicated (Ref. F-13).

The Mattauch-Herzog mass spectrometer (Section 2.3.1.2.2) was part of the SPAS-01 subsatellite which was detached from the Shuttle on STS-7 and flew up to 330 m behind the Shuttle (Ref. F-14). The sampling rate was one point every 2 sec for 28 amu (N_2 + CO) and one point every 4 sec for 32 amu (O_2). At point "a" of Figure 2.3-25 the subsatellite was detached from the Shuttle. In the period "b" the attitude was changed such that the ion source was in the ram direction. From then on the signals were due to ambient molecular nitrogen and to ambient atomic oxygen recombining in the ion source to molecular oxygen. During the period "c" the shuttle moved toward the SPAS and the "grab" occurred at "d". Frequent firings of the thrusters of the Shuttle created N_2, CO, and H_2O

which got into the vicinity of the subsatellite. The water vapor deposition on the ion source surfaces hindered the recombination of atomic oxygen into O_2, which appeared to give an O_2 signal negatively correlated with the nitrogen pulses.

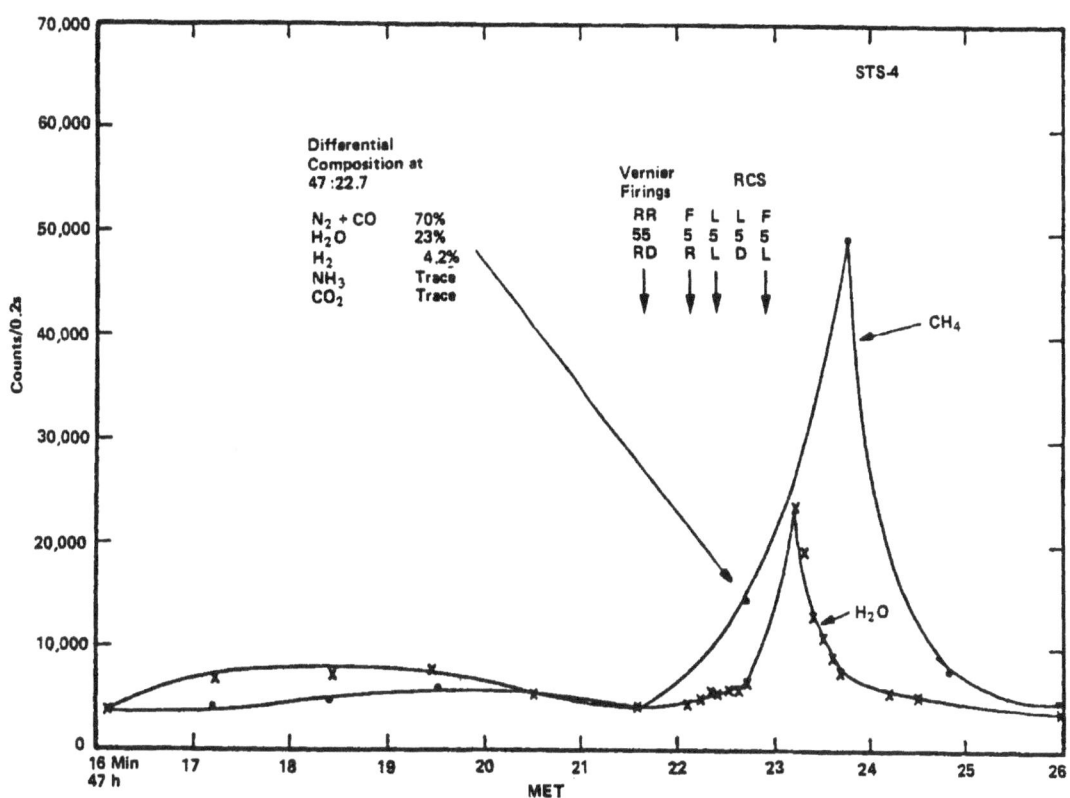

Figure 2.3-23
Signature of a Vernier RCS Firing during Mapping (Ref F-12)

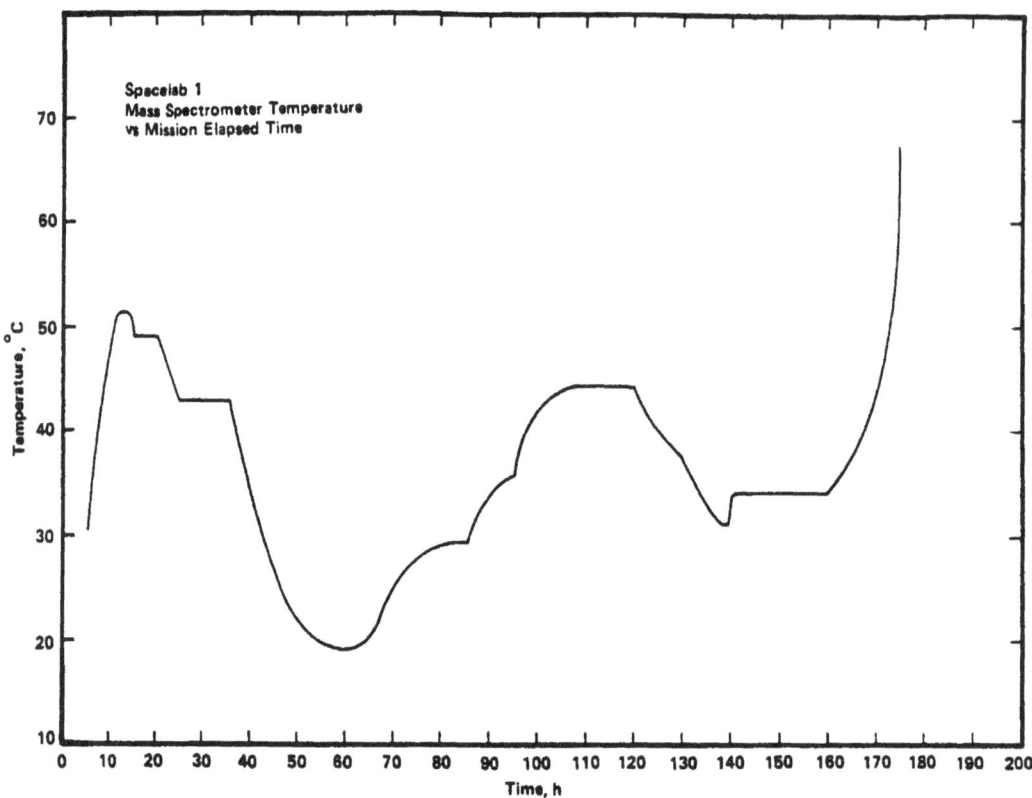

Figure 2.3-24 Mass Spectrometer Temperature Versus Mission Elapsed Time (Ref F-13)

Figure 2.3-26 shows the mass spectrum obtained on STS-11 during the sixth day inside the payload bay with the ion source facing the forward bulkhead. The most prominent contaminant is water vapor at 18 amu. If the contaminant at 19 amu is H_3O it is not yet certain whether it was produced within the halo of contamination surrounding the Shuttle or inside the ion source of the instrument (Ref. F-14). Mass 40 is most likely argon, and mass 44 is CO_2. The peaks at masses 67 and 69 cannot be identified as hydrocarbons because of lack of other peaks on either side of this pair of peaks. Certain Freons produce peaks in the 67 to 69 mass region but again more peaks on either side would be expected. (Note above that on STS-4, Freon was detected in the vicinity of the aft bulkhead).

2-102

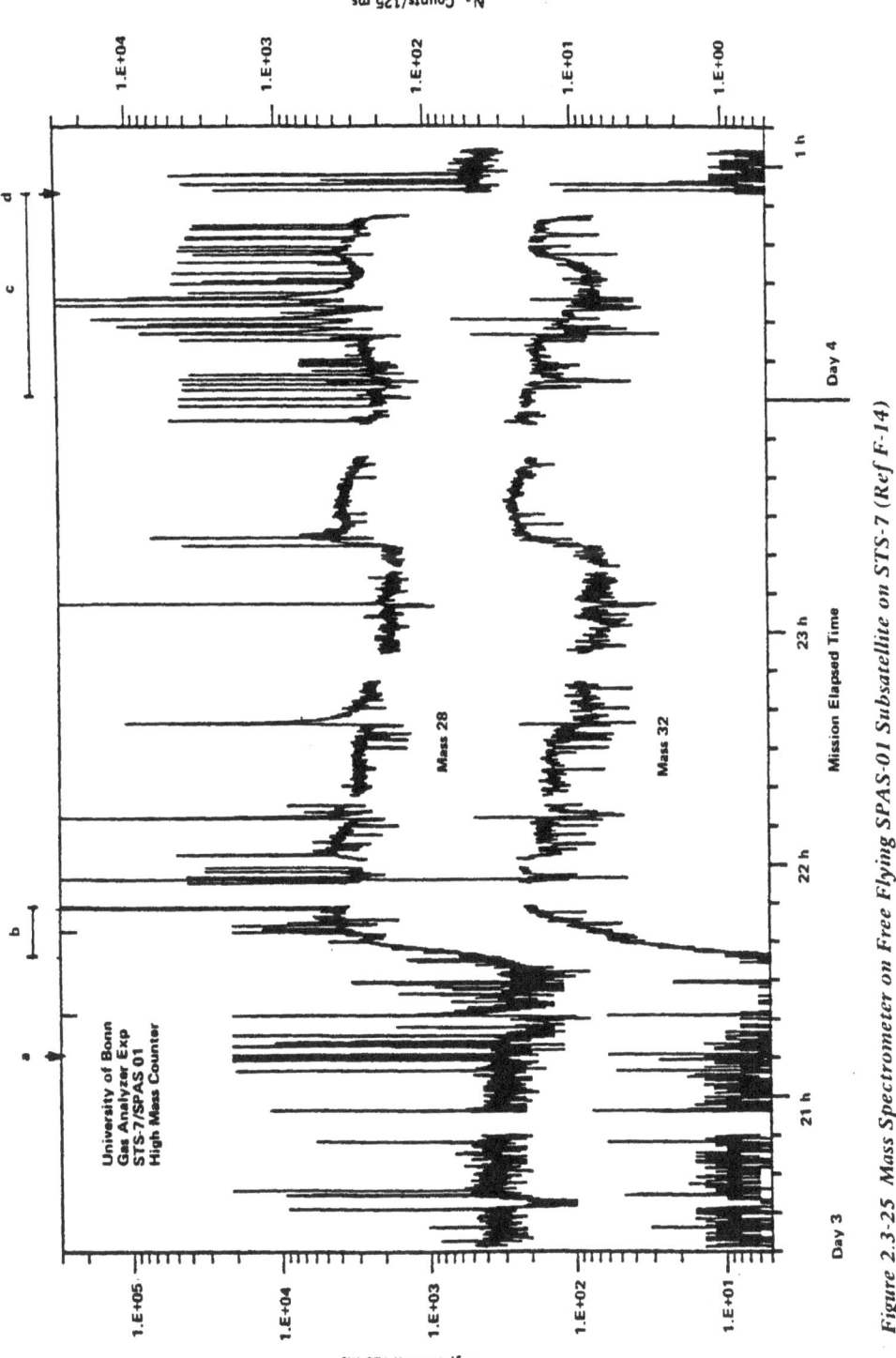

Figure 2.3-25 Mass Spectrometer on Free Flying SPAS-01 Subsatellite on STS-7 (Ref F-14)

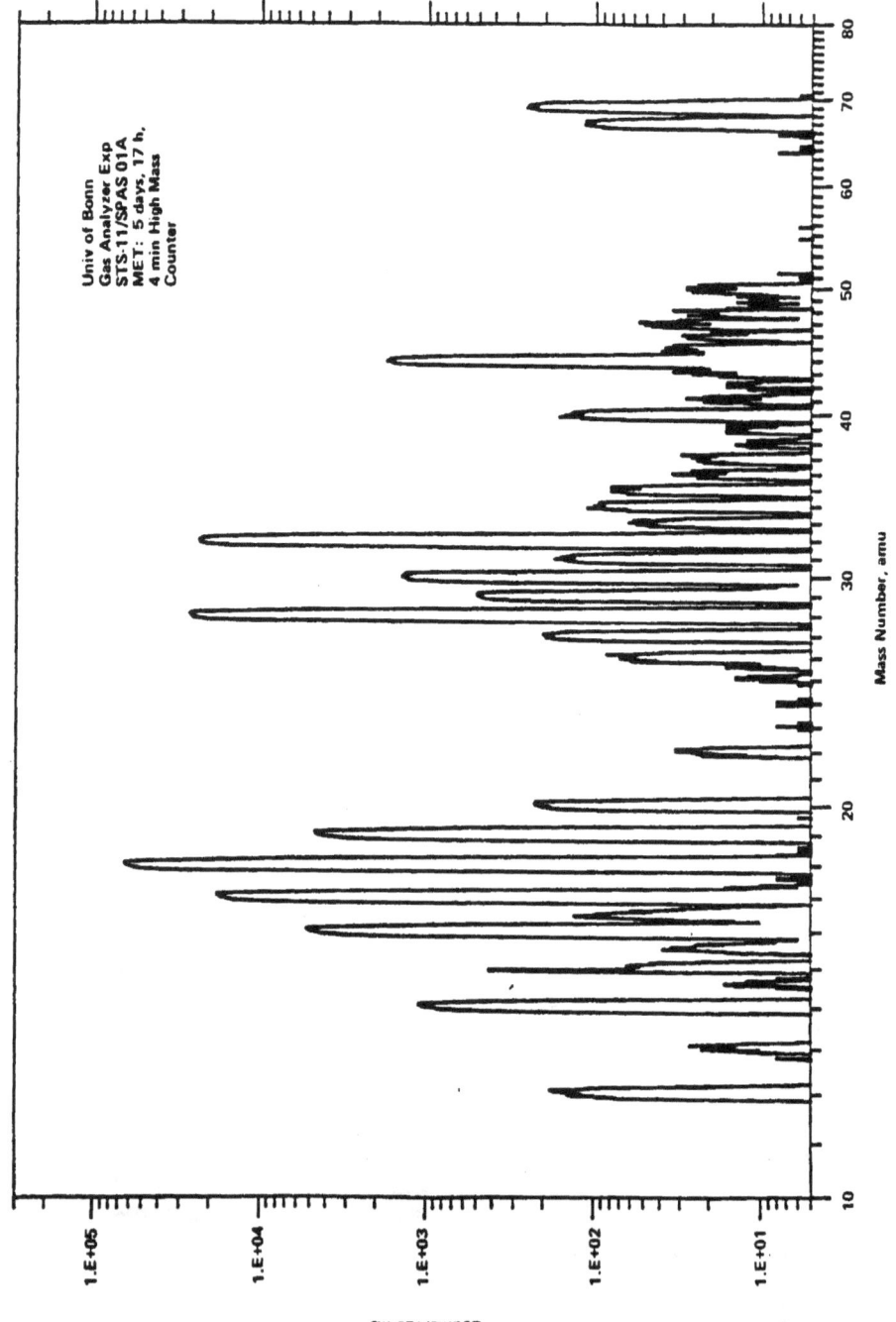

Figure 2.3-26 Mass Spectrum of Payload Bay on STS-11 (Ref F-14)

Figure 2.3-27 shows the mass spectrum of the bay during astronaut activity in the bay. The effects of multiple thruster firings of the Manned Maneuvering Unit (MMU) are very clearly seen on the total ion current electrometer (upper trace). The average level of contaminants was so high that on the lower trace the readings are from the d.c. electrometer because use of the multipliers saturated the signal for many lines. It can be noted from the two traces that the contamination consisted of a quickly variable component and a slowly varying one. The TIC electrometer (upper trace) shows that the majority of all emitted gases disappear with time constants of less than a second. The slowly varying component persists after the MMU thruster firings and is dominated by water vapor and by CO at 28 amu.

Comparison of Figure 2.3-18 with Figure 2.3-27 shows the Mattauch Herzog spectrometer to be much more successful for mass resolution than the quadrapole instrument of the IECM.

2.3.3.1.3 Descent

During descent on STS-2, -3, and -4 the Humidity Monitor (IECM01) recorded the relative humidity in the cargo bay. The results are depicted in Figure 2.3-28 (Ref. F-15).

The Air Sampler collection timeline during descent of STS-2 is depicted in Figure 2.3-29, and the analysis of the collected contaminants are listed in Table 2.3-11 (Ref. F-1). As with the contaminants of the Air Sampler during ascent (Sec 2.3.3.1.1) analysis was made by gas chromatograph/mass spectroscopy. The gases identified in analysis of the samples on STS-9/SL-1 are listed in Table 2.3-12 (Ref. F-3).

The mass accumulation by the TQCM (IECM07) as measured on STS-2 and ST-4 are listed in Table 2.3-13 (Ref. F-4) (No descent data was taken on STS-3). Figure 2.3-30 shows a representative plot of mass accumulation during the entire descent phase for the +Y axis sensors and their temperatures. On STS-2 the sensors were in the +80°C cleanup mode when the de-orbit command occurred. On STS-4 they were at -30°C at de-orbit command. An increase in mass accumulation is seen at landing with a slow increase after landing until IECM power-down.

The data for the two CQCM (IECM08) sensors for each of the three flights with the exception of the -Z1 sensor on STS-4 are presented in Table 2.3-14 (Ref. F-4). The -Z1 on STS-4 showed extreme sensitivity to direct solar radiation causing the data to be unreliable.

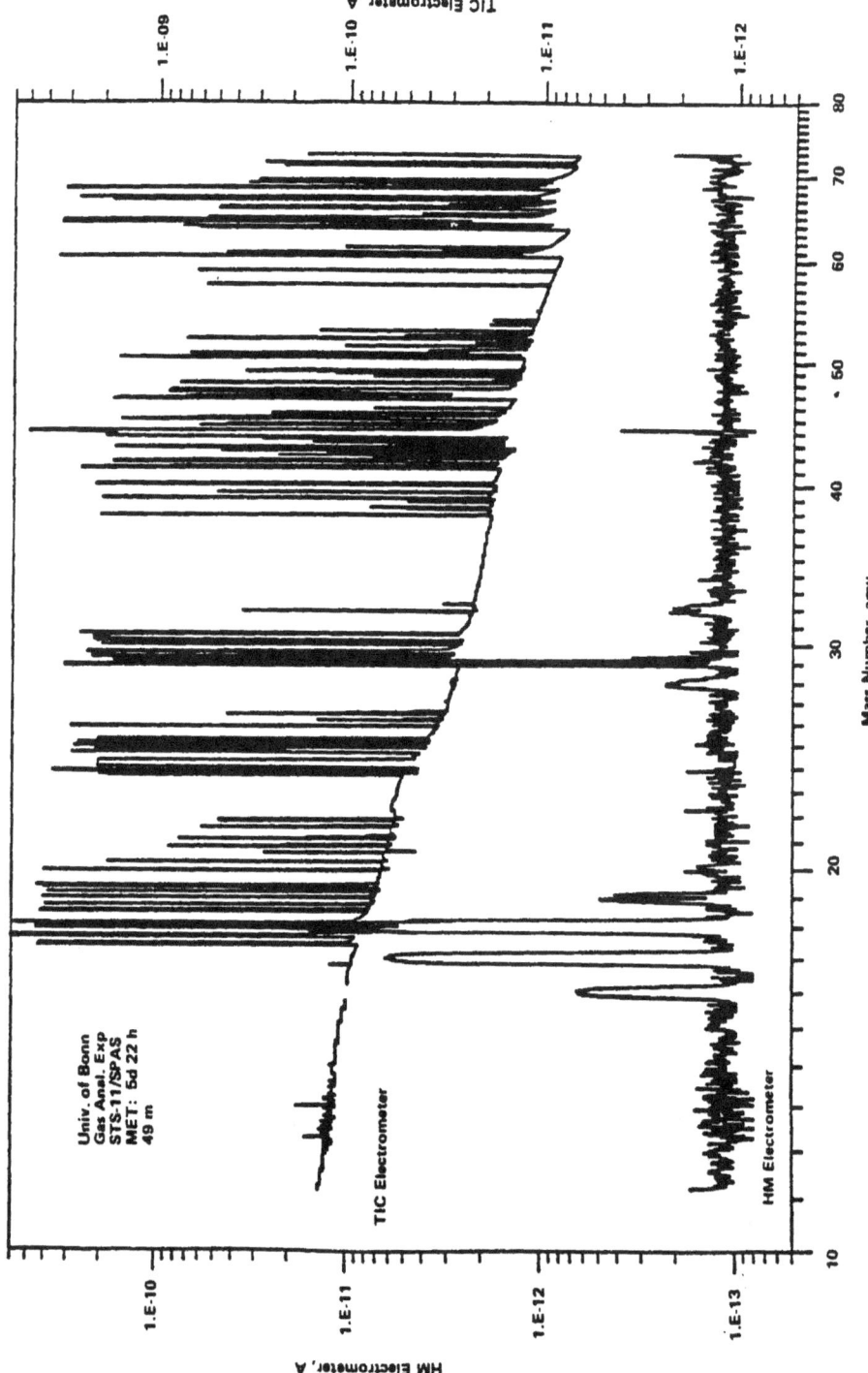

Figure 2.3-27 Mass Spectra of the Payload Bay during Astronaut Activity in the Bay (Ref F-14)

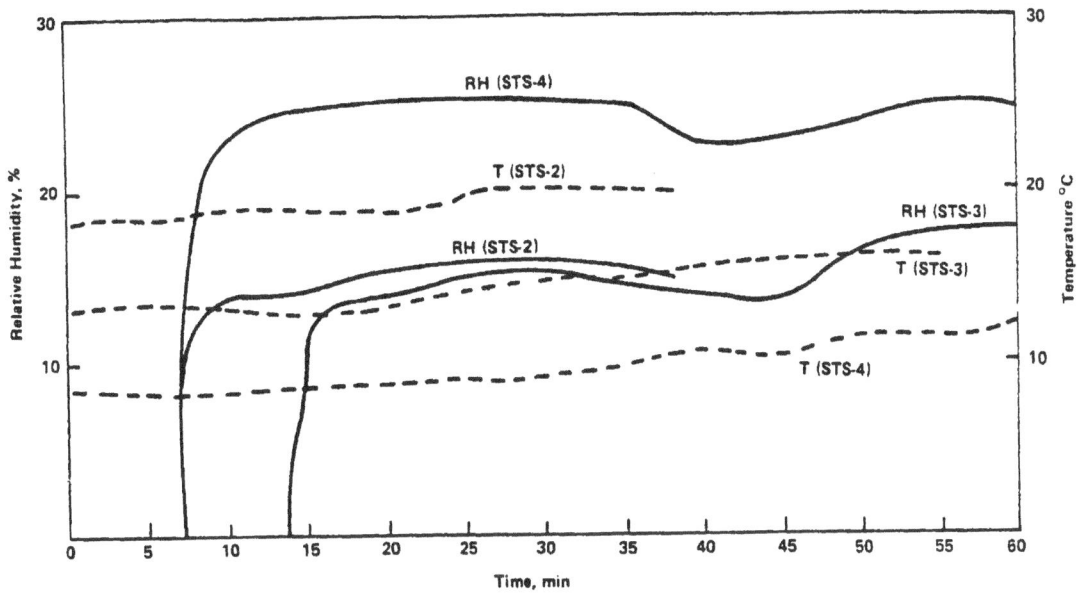

Figure 2.3-28
IECM Reentry Humidity Monitor and Reentry Temperature (Air Sampler) for STS-2, STS-3, and STS-4 (Ref F-15)

Table 2.3-11
IECM Air Sampler Organics Detected in Most Significant Quantities* on STS-2 (Ref F-1)

Compound	Descent B, μ gm
$C_9H_{16}O$ (2-Isononenal)	36.0
4-Methyl-1-Pentene	79.0
1, 1, 2-Trichloro-1, 2, 2-Trifloroethane	22.0
Methyl Benzene	11.0
Dibromochloromethane	8.8
Hexamethylcyclotrisiloxane or Similar Compound	30.0
A Carboxylic Acid Ester, No Satisfactory Match	—
Benzenedicarboxylic Acid, Di C_{14} Ester	5.1
Nonadecane	2.5
1, 2-Benzenedicarboxylic Acid, Diethyl Ester	20.0
Aromatic Hydrocarbon (Possibly Oxygenated)	10.0
	In Approx 315 Std CC

*Subsequent control analyses show that these quantities must be lowered significantly.

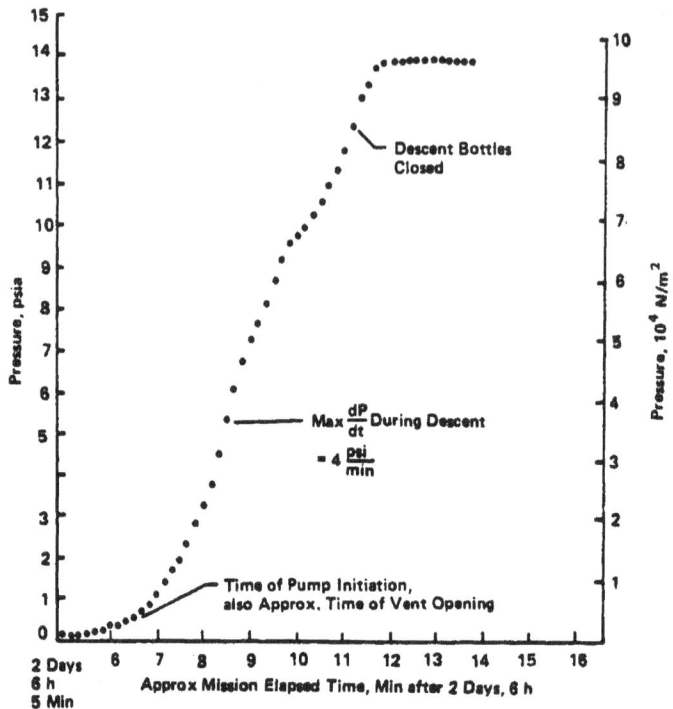

Figure 2.3-29 Pressure During Descent of STS-2 (Ref F-1)

Table 2.3-13 Mass Accumulation by TQCM during the Descent Phase (Ref F-4)

STS-2 Descent Phase Total Duration: 1h, 10 min								
Mission Elapsed Time, h:min	Sensor Axis	53:50	54:00	54:10	54:20	54:30	54:40	54:44
Mass Accumulation per Area, ng cm^{-2} Sensor Temperature, °C	+Y	0 18.0	55 17.1	58 16.2	123 16.2	140 16.2	150 17.1	150 17.1
Mass Accumulation per Area, ng cm^{-2} Sensor Temperature, °C	+X	0 16.2	78 15.2	80 14.3	374 14.3	395 14.3	413 14.3	413 14.3
Mass Accumulation per Area, ng cm^{-2} Sensor Temperature, °C	-X	0 17.1	56.2 17.1	44 17.1	162 19.0	136 20.8	125 21.8	125 21.8
Mass Accumulation per Area, ng cm^{-2} Sensor Temperature, °C	-Y	0 15.2	9.4 14.3	-36 13.3	-75 17.1	-115 19.0	-92 20.8	-92 20.8
STS-4 Descent Phase Total Duration: 1 h, 28 min								
Mission Elapsed Time, h:min	Sensor Axis	168:45	169:00	169:15	169:30	169:45	170:00	
Mass Accumulation per Area, ng cm^{-2} Sensor Temperature, °C	+Y	0 -1.6	62 -2.5	137 3.1	165 7.7	182 10.5	179.4 +13.5	
Mass Accumulation per Area, ng cm^{-2} Sensor Temperature, °C	+X	0 5.9	-19 +4.9	0 +4.9	16 +5.9	8 6.8	8 8.8	
Mass Accumulation per Area, ng cm^{-2} Sensor Temperature, °C	-Z	0 6.8	-13 6.8	343 5.9	360 6.8	349 8.7	360 9.6	
Mass Accumulation per Area, ng cm^{-2} Sensor Temperature, °C	-X	0 7.7	4.7 6.8	190 10.5	195 13.3	208 15.2	220 16.2	
Mass Accumulation per Area, ng cm^{-2} Sensor Temperature, °C	-Y	0 4.0	2 3.1	-183 7.7	-206 10.5	-192 13.3	-178 14.3	

Table 2.3-12 Species from Descent Air Sampler on STS-9/Spacelab 1 (Ref F-3)

Peak Number	Area Percent	Amount in 10^{-9} g per Substance	Substance Identity
1	0.6	21	Unidentified
2	2.5	87	Branched Alkane
3	0.2	7	Branched Alkane
4	0.9	32	Branched Alkane
5	1.3	46	C_7 Branched Alkane
6	0.6	21	Branched Alkane
7	0.6	21	Branched Alkane
8	0.8	28	Unidentified
9	1.4	49	C_9 Branched Alkane
10	2.7	93	Octamethylcylotetrasiloxane
11	0.7	25	Branched Alkane
12	0.7	25	Branched Alkane
13	0.7	25	Branched Alkane
14	0.5	18	Branched Alkane
15	3.0	105	Branched Alkane
16	0.3	10	Branched Alkane
17	2.5	88	Branched Alkane
18	1.4	49	Branched Alkane
19	1.6	56	Branched Alkane
20	1.6	56	Branched Alkane
21	0.9	32	Unidentified
22	0.7	25	Branched Alkane
23	0.6	21	Unidentified
24	0.6	21	Tridecane
25	0.6	21	Ester or Diester
26	0.3	11	Unidentified
27	4.7	165	Phthalate Isomer (Plasticizer)
28	0.9	32	Branched Alkane
29	4.1	143	Phthalate Isomer (Plasticizer)
30	0.6	21	Unidentified
31	1.6	56	Branched Alkane
32	0.7	25	Anthracene
33	1.0	35	Carboxylic Acid
34	1.7	60	Branched Alkane
35	0.8	29	Unidentified
36	0.7	25	Branched Alkane
37	1.0	35	Heptadecane
38	1.4	49	Branched Alkane
39	3.8	133	Phthalate (Dibutyl?)
40	1.3	46	Unidentified
41	6.5	228	Unidentified
42	1.9	67	Unidentified
43	0.9	32	Branched Alkane
44	0.3	16	Branched Alkane
45	0.7	25	Branched Alkane
46	0.8	29	Unidentified
47	3.2	112	Branched Alkane
48	0.3	11	Branched Alkane
49	0.5	16	Unidentified
50	3.8	133	Alkane (Branched ?)
51	0.4	14	Branched Alkane
52	4.1	144	Alkane (Branched ?)
53	6.4	224	Alkane (Branched ?)
54	8.8	308	Alkane (Branched ?)
55	8.6	301	Alkane (Branched ?)
56	0.2	7	Branched Alkane

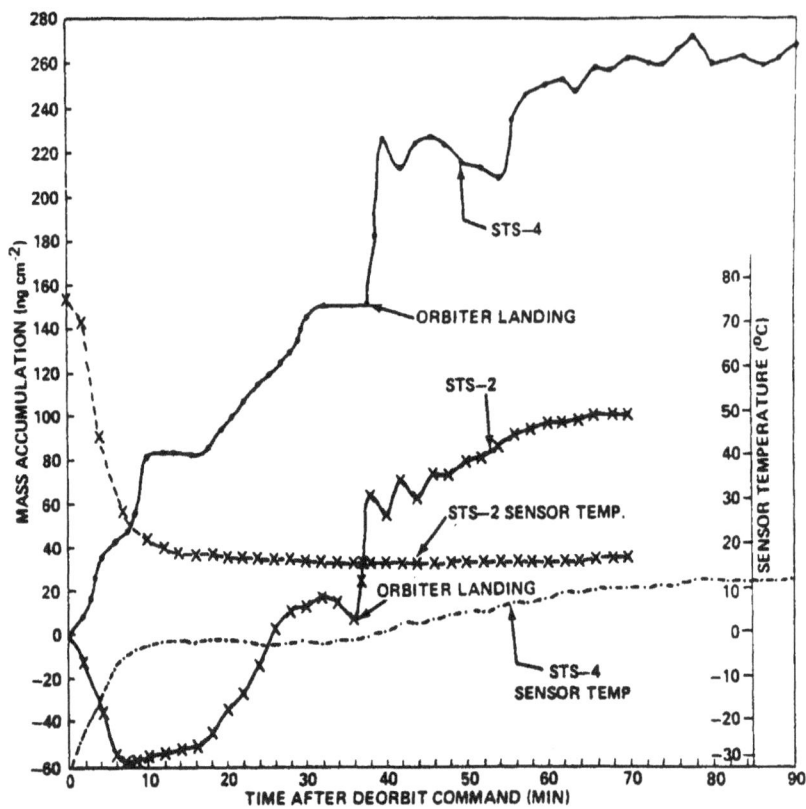

Figure 2.3-30
Mass Accumulation during Descent Phase on +Y Axis TQCM Sensors on STS-2 and STS-4 (Ref F-4)

Table 2.3-14 CQCM Net Molecular Mass Accumulations Rates during Descent Phase (Ref F-4)

Attitude	MET, h min	Δ Time, min	Sensor -Z1 Sensor Temp, °C	Mass Change, ng cm^{-2}	Mass Accum Rate, ng cm^{-2}/h	Sensor -Z2 Sensor Temp, °C	Mass Change, ng cm^{-2}	Mass Accum Rate, ng cm^{-2}/h
STS-2	053 34 / 054 45	71	4 / 27	78	66	4 / 26	112	95
STS-3	191 29 / 192 59	90	14 / 25	0	0	23 / 25	37	25
STS-4	168 30 / 172 02	92				-4 / 23	-41	-26

2.3.3.2   Additional Flight Information

2.3.3.2.1   Unpublished Flight Molecular Information

The IRT of SL-2 (STS-51F) provides some information on the molecular environment on that Spacelab flight. Flight data has not been completely evaluated at the time of this handbook edition. The instrument experienced some failures, including the presense of some Mylar insulation in the field-of-view of the sensor during at least part of the mission. With this in mind, the contamination implications of some of the IRT data from early in the flight has been examined by Dr. Fred Witteborn, NASA ARC.

Preliminary evaluation of a very small amount of data early in the mission is consistent (on 3 of 6 channels) with water column densities of $3 \times 10^{13}/cm^2$ (and higher densities on the other 3), and a density of CO_2 of $1.3 \times 10^{13}/cm^2$ on a fourth. Whether or not the mylar insulation was present, and what effect this material would have on the IR signal are currently unknown. It is hoped that a more detailed evaluation of the data will illuminate these issues.

One other set of unpublished data was collected by coated mirror samples flown on the Hitchhiker payload carrier flown on STS-61C.

2.3.3.2.2   Needed Flight Molecular Information

As will be discussed in Section 2.3.3.3.2, the effects of deposited material on surfaces can very greatly, depending on the composition (and morphology) of the deposit. Attempts to measure the effects of flight deposits (the OEM of the IECM) thus far have been unsuccessful, due to the small deposit thicknesses collected. A desirable set of flight data would be a measure of the effects of deposited molecules during the vari- ous classes of Shuttle Flights. Another less direct approach to measuring deposition effects would be to include volatile condensible material (VCM) effects in a standard outgassing materials test. This approach would required that extensive analyses be conducted to predict flight degradations.

2.3.3.2.3   Planned Flight Molecular Information

At this time, no dedicated molecular contamination experiments planned for flight are known. The USAF IOCM may fly again, but data collected by this instrument may continue to be classified and unavailable. The same may be true of data collected by CIRRUS and other USAF or Strategic Defense Initiative Office (SDIO) infrared instruments. Thus, release of molecular data from future flights can not be guaranteed.

2.3.3.3   Flight Molecular Tools/Models

2.3.3.3.1   SPACE II Computer Model

Spacecraft external nonmetallic materials under the influence of the vacuum environment of space inherently demonstrate a loss of mass characteristic of the particular material. Operational systems such as attitude control engines, overboard vents and pressurized compartments can emit copious amounts of contaminant material to space vacuum. The combination of all such phenomena for a particular space vehicle produces a dynamic molecular and particulate contaminant environment in the near vicinity of the vehicle.

Previous methods of evaluating this complex phenomena were limited to isolated analyses of simplified spacecraft configurations requiring numerous simplifying assumptions and computer modeling approaches dealing with only specific aspects of the total problem. These approaches were practical for past space programs which demonstrated limited variations in the major influencing parameters. With the inception of the Shuttle/Spacelab Program presenting almost unlimited variations in influencing parameters (both mission and configuration dependent), utilization of earlier simplified approaches became untenable. It became apparent in attempting to evaluate the Spacelab vehicles for compliance with program on-orbit contamination control criteria and in conducting Spacelab/Shuttle Orbiter mission feasibility analyses for MSFC that the need existed for an all-up systems level contamination evaluation computer model to handle the enormous number of calculations required to accurately predict the total induced molecular environment. The resulting Shuttle/Payload Contamination Evaluation Version II computer model (denoted the SPACE II Program) was developed with the prime objectives of refining the modeling approaches to describing the complex physical phenomena involved and integrating previously developed methodology into a coherent systems level computer program capable of accommodating all primary variables to dynamically simulate the molecular induced environment.

The Martin Marietta developed SPACE II Program is a systems level computer model which mathematically synthesizes and maps the induced molecular contaminant environment of the Space Shuttle Orbiter and the Spacelab carrier vehicles to be flown as payloads within the Orbiter bay. SPACE II has been configured to accept other space vehicles for evaluation as well. The purpose of the SPACE II Program is to simulate the complex dynamics of a spacecraft's induced environment to establish the levels of surface deposition and contaminant cloud thickness which will impact the operation of spacecraft systems and degrade scientific data acquired by sensitive instruments. Development of the SPACE II Program was prompted by the need to minimize manual calculational requirements by integrating numerous independent subprograms and analytical approaches into an all-encompassing model which could simultaneously consider vehicle configuration, all major contaminant sources and their interactions and the transport of these sources to spacecraft/instrument surfaces or to locations within the volume of space through which scientific instruments would view (i.e., a total systems level model).

Prior to the SPACE II Program development and subprogram integration, the analytical approach and methodology of many of the component subprograms were verified by flight data obtained during previous manned and unmanned space programs. Deposition prediction subroutines for direct line-of-sight contaminant transport from source to surface were validated during the Skylab Program for an assortment of external experiment surfaces (such as the D024 Thermal Coating Experiment and the T027 Optical

Contamination Experiment), Skylab windows, solar arrays and on-orbit deposition detectors. For example, Figure 2.3-31 presents the premission deposition predictions for one of the six active Skylab Quartz Crystal Microbalance (QCM) deposition monitors and the actual in-situ deposition levels measured during the Skylab Program (Ref. F-16). The premission predictions were so close to the as-flown flight data that only minor "fine tuning" (of outgassing decay rates) was required for near exact correlation.

Subroutines developed to determine the backscatter of emitted contaminant molecules resulting from collisions with the on-orbit ambient atmosphere (denoted as return flux) were verified for accuracy with a neon gas experiment flown on the Atmospheric Explorer-D Satellite. This experiment expelled a known amount of neon gas into the atmospheric "wind" and the amount returned to the satellite was detected by a mass spectrometer. The SPACE II Program return flux subroutine was utilized to predict the return flux levels based upon the satellite configuration, the orbital conditions and the neon vent flowrate and plume geometry. Figure 2.3-32 demonstrates the close correlation obtained between predicted and measured return flux levels while the neon experiment valve was open (Ref. F-17).

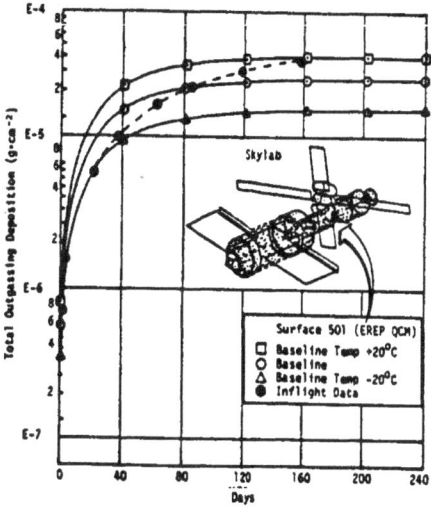

Figure 2.3-31
Skylab Premission Prediction Correlation with Flight Data (Ref F-16)

In addition to the above mentioned space programs, various subroutines and versions of the SPACE II Computer Program have been applied to evaluations of NOAA-ITOS satellites, the DSP and DMSP satellites, the SCATHA Program, Titan launch vehicles, over 20 Shuttle DOD payloads and is currently being applied to programs such as Magellan and Space Station. This demonstrates that the SPACE II Program has not only been used extensively but is also a highly flexible analytic tool adaptable to numerous complex situations. Recently the SPACE II Program has been verified for accuracy through comparison of Shuttle/Spacelab IECM flight data with SPACE II Program premission contamination predictions for the QCM sensors and mass spectrometer (Ref. F-18 and F-19). Integrated deposition predictions were found to be accurate, although the model is not able to predict short term variations, such as those that occur in a single orbit.

The SPACE II Program is written completely in FORTRAN IV and is currently operational on the Martin Marietta Aerospace CDC Cyber 750 and VAX computer systems, the UNIVAC 1108 system at NASA's Marshall Space Flight Center and the UNIVAC 1110 system at Johnson Space Center. The amount of core required depends on the machine, the operating system and the efficiency of its compiler and loader.

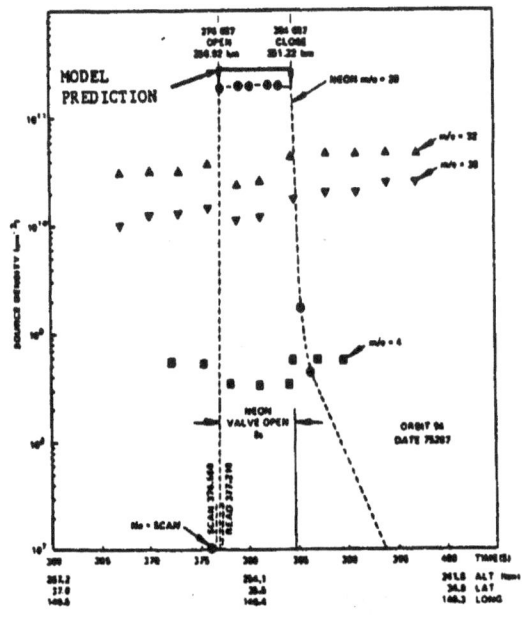

Figure 2.3-32
Atmospheric Explorer-D Return Flux Experiment Results (Ref F-17)

Core requirements for the SPACE II Program have been minimized through extensive overlay structuring of the model subroutines which are called into the main program only when required for a specific calculation. Random mass storage is utilized in the manipulation of the SPACE II Program input data to minimize model run time which is characteristically less than a few hundred seconds for a typical run.

A comprehensive description of the SPACE II Program is contained in the SPACE Program User's Manual, (Ref. F-20). Copies of this document are available at the repository at MSFC. This section contains only a summary of the program operation, function and methodology. Contained therein are descriptions of the modeled configurations, contaminant source transport relationships, program logic flow, subroutines and permanent data files. User input and output options are discussed and sample problems are presented. In addition, the User's Manual contains complete documentation into the methodology, physics and assumptions utilized in the development of the various SPACE II Program subroutines.

The SPACE II Program, as delivered to the aforementioned NASA centers, has the capability of evaluating the induced contaminant environment of the Space Shuttle Orbiter and three representative Spacelab configurations. It is configured, however, with the ability to evaluate any arbitrary vehicle configuration through proper input data manipulation. SPACE II considers all major Spacelab and Orbiter contaminant sources including external nonmetallic materials outgassing and early desorption, leakage from pressurized crew compartments, overboard vents such as the Orbiter water evaporator vents and attitude control engines including the six VRCS and the thirty-eight PRCS monomethyl hydrazine/nitrogen tetroxide hypergolic engines. These sources are modeled as closed form mathematical expressions describing source emission rates and emission patterns based upon available nonmetallic materials vacuum test data, in-situ testing of engine and vent systems and engineering analysis. New or modified sources can be easily added to the existing configurations or to any new arbitrary vehicle configuration developed.

Once the molecular contaminants have been emitted from a particular source, their transport to spacecraft/payload surfaces or to locations within the hemispherical volume above the spacecraft are simulated within the SPACE II Program (see Figure 2.3-33). The general expression used to describe the percentage of emitted source material (X) capable of reaching location (Y) is denoted as the mass transport factor (MTF) from X to Y or MTF_{X-Y}. The MTF is a function of not only the vehicle geometry but also the emission characteristics of the specific source. The contaminant transport phenomena considered in the SPACE II Program include the following (refer to Figure 2.3-33 for examples):

  a. direct source to surface (① to ③ and ⑥ to ⑦) or to a location in space (② to ④);
  b. direct source to surface or to a location in space with attenuation due to molecular collisions with the ambient atmosphere ⑤ prior to reaching the location of interest ④;
  c. reflection/reemission from vehicle surfaces ⑦ or ③ —
     ($\dot{m}_{ev}$);

d. return flux resulting from collisions with the ambient atmosphere (④ to ③) and
e. return flux resulting from collisions with other contaminant species (self-scattering).

The primary output of the SPACE II Program includes the mass column density (MCD in g/cm^2) or the molecular number column density (NCD in molecules/cm^2) along instrument lines-of-sight, the return flux levels to surfaces of interest (molecules/cm^2s) and accumulative surface deposition levels (molecules/cm^2) for the contaminant sources and transport phenomena evaluated.

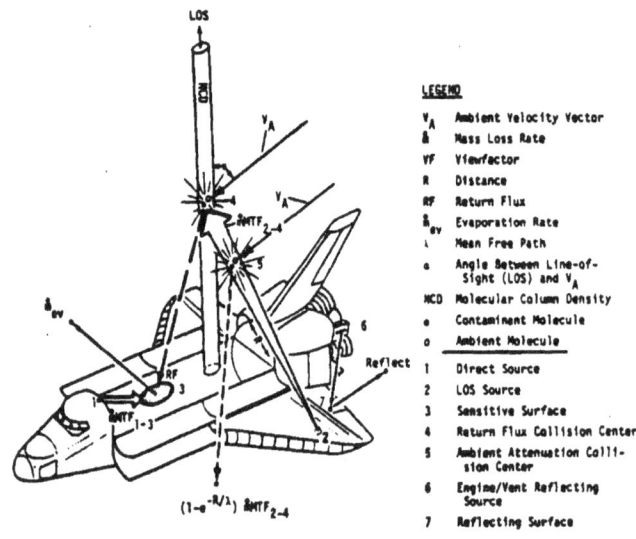

Figure 2.3-33
Illustration of SPACE II Contaminant Transport Functions

Output from the SPACE II Program can be used as input to separate or independent subroutines and analytical approaches to determine the degradation effects induced by the predicted contaminant levels of surface deposition and contaminant cloud thickness (NCD). As a result, predictions can be made for such effects as surface reflectance or transmission loss and radiant scattering, emission or absorption by the molecular cloud.

Execution of a SPACE II Program contamination prediction run is controlled by the program executive which is the primary mechanism by which the separate contamination analysis subroutines were integrated into a complete systems level model capable of simultaneous evaluation of all major phenomena influencing contamination. The program executive through true/false control flag manipulation directs input data flow, block data modification and the call of required model segments into core. The model has been configured with preset "default" input which allows the user an easy means of program checkout prior to more detailed computer runs.

Under the control of the program executive, data from permanent files or tape are called into the run stream. These files contain the required configuration data and corresponding surface temperature data for the analysis being conducted. Input configuration data (which is basically a permanent file of the mass transport factors between all sources and locations of interest, is developed through use of a modified Thermal Radiation Analysis System (TRASYS, Ref. F-21) black body radiation program. TRASYS calculates the MTF (i.e., the percentage of mass emitted by a Lambertian source surface capable of impinging upon another surface or location in space) and geometrical relationships such as source/surface separation distances and angular relationships used in SPACE II to establish contaminant impingment from non-Lambertian source emission patterns. The mass transport factor data files for the Orbiter and Spacelab configurations have been precalculated and are addressable from permanent data files. Modifications of Spacelab or Orbiter geometry or development of a new vehicle configuration requires creation of new or modified mass transport factor data files through the TRASYS Program. Vehicle configuration input to SPACE are developed by inputting appropriate geometrical shapes (i.e., cones, cylinders, planes, etc.) into TRASYS. The TRASYS program calculates the formatted mass transport factors with surface shadowing included, develops hard copy plots of the integrated configuration for visual verification of input data and assigns each modeled surface node a unique surface identification number. The assigned surface numbers serve as the basic library indexing system by which the SPACE II Program designates source characteristics and temperature profiles to the modeled surfaces.

2.3.3.3.2 Other Analysis Tools

Simple techniques for estimating self scattered and ambient scattered return flux have been developed (Ref. F-22). These techniques do not offer the accuracy of the computer calculations previously described.

The return fluxes ϕ_r, produced by the scattering of emitted fluxes ϕ_d with ambient molecules can be estimated from

$$\phi_r = \phi_d(R/\lambda_0)(V_s/V_d + 1) \sim 21(R/\lambda_0)\phi_d, \qquad 2.3\text{-}7$$

where
- R = Hemisphere radius, assumed for Shuttle bay ~2.4 m;
- λ_0 = molecular mean free path;
- V_s = orbit velocity ~8 km/s; and
- V_d = average velocity of emitted molecules ~0.4 km/s.

For nitrogen, the return fluxes out of the total direct flux for various altitudes are shown in Figure 2.3-34. The return flux of water molecules at 241 km has been calculated to be about $2 \times 10^{13}/cm^2 s$ at 2.35 MET and about $1 \times 10^{11}/cm^2 s$ at 150 MET. These compare to the water initial return flux of 10^{12} minimum to 10^{14} maximum and fluxes of 10^{11} to 10^{13} at the end of the flight for STS-2, -3, and -4 (Ref. F-2).

The scattering of the outgassed molecules among themselves is about three orders of magnitude less than the direct flux and less than the return flux at 241 km orbit. It can be estimated from

$$\phi_{ss} = 1.78 \times 10^{-2} (\sigma R/V_d) \phi_d^2 / cm^2 s \qquad 2.3\text{-}8$$

where σ (cm^2) is the average cross section of the outgassing molecules and the other symbols are as defined above.

The molecular column density NCD ($/cm^2$) or the mass column density MCD (g/cm^2), representing the number of molecules or the mass of the molecules (MCD = A_nNCD, where A_n is Avogadro's number) in a column of 1 cm^2 extending from the bay to infinity can be estimated from

$$N_c = (\lambda_0/V_s)\phi_r \sim (R/V_s)\phi_d \sim nR/cm^2 \qquad 2.3\text{-}9$$

where n = density ($/cm^3$).

The baseline column density for the Shuttle from measurements of pressure in the bay are depicted in Figure 2.3-35. A column density of less than $10^{12}/cm^2$ water molecules occurs at 3 to 4 hours MET which is obtained when considering the water to be about 3 percent of the total column. The column densities in Figure 2.3-35 agree with the measured maximum and minimum values of 3×10^{13} and $4 \times 10^{10}/cm^2$ found for STS-2, -3, and -4 (Ref. F-2).

2.3.3.3.3 Molecular Contamination Effects Models

The effects of molecular contaminants are almost completely dependent on the type of molecules involved. For deposited molecules, the optical properties of the contaminated surface will degrade toward those of the deposit. These properties (solar absorptivity, reflectivity, transmissivity, emmissivity, etc.) can vary greatly among different substrates and contaminants. Prediction of degradation of specific materials by deposited molecular contaminants depends on accurate prediction of the contaminant(s) deposited. Predictions of the material deposited are in most cases extremely difficult to make, and the techniques for making these predictions are generally beyond the current state-of-the-art. If an instrument is determined to be sensitive to organic or silicone deposits (assuming the deposit consists of some representative species) of the thicknesses indicated by the flight of data of Section 2.3.3.1, then protection techniques should be considered.

Prediction of the effects of field-of-view contaminants are also very complex. The molecules most likely to appear in fields-of-view around the Orbiter are N_2, O_2, H_2O and CO_2. Reference F-23 describes the calculation of received power from field-of-view H_2O and CO_2 molecules around the Shuttle. Similar techniques could also be applied to other species.

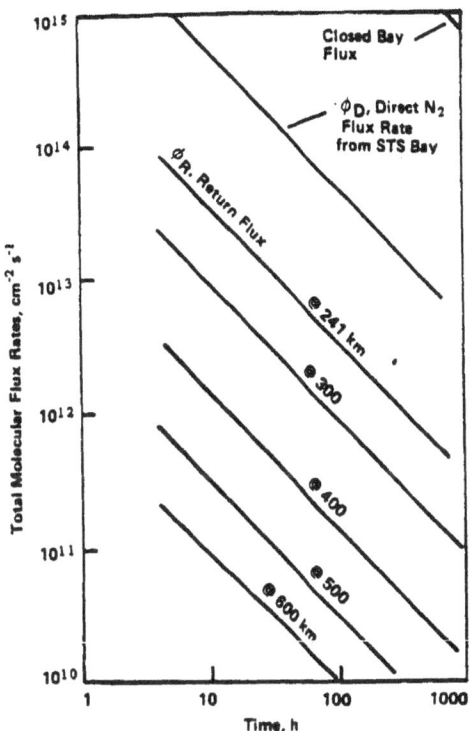

*Figure 2.3-34
Total Direct and Return Flux vs Time in Baseline
STS Bay—Standard Atmosphere*

2.3.3.4 <u>Contamination Prevention Techniques</u>

Analysis has shown that a positive pressure and purging flow by clean gas inside instrumentation can prevent ingestion of external molecular and self-generated (offgassing) molecular contaminants. The following is extracted from Reference F-24.

Insufficient venting and improper locations may subject a compartment to large pressure differentials and structural failures. An approach to venting of a compartment is to have a flow response time comparable or faster than the external flow field disturbance time. Assuming isothermal flow conditions and small pressure differentials the volume V venting through an orifice, A, and a discharge coefficient, C_D, is $d(\rho V)/dt = \rho v A C_D$ where ρ is the gas density and v its velocity at the orifice. From the gas law $\rho = P/RT$ and its derivative $d\rho/dt = (1/RT) \, dP/dt$. The gas velocity is

$$v = (2gh)^{1/2} = [2g(\Delta P/\rho)]^{1/2} = [2gRT(\Delta P/P)]^{1/2}. \qquad 2.3\text{-}7$$

Substitution, with V, A, and C_D as constants, leads to

$$\frac{dP}{dt} = \frac{AC_D P}{V} [2gRT(\Delta P/P)]^{1/2} \qquad 2.3\text{-}8$$

and

$$\Delta P = \frac{1}{2gRT} \left[\frac{V^2}{AC_D}\right] \frac{(dP/dt)^2}{P} \qquad 2.3\text{-}9$$

therefore (with $dP/dt = \dot{P}$)

$$\frac{V}{AC_D} = [2gRT(P_o[\Delta P/\dot{P}^2])] \, m. \qquad 2.3\text{-}10$$

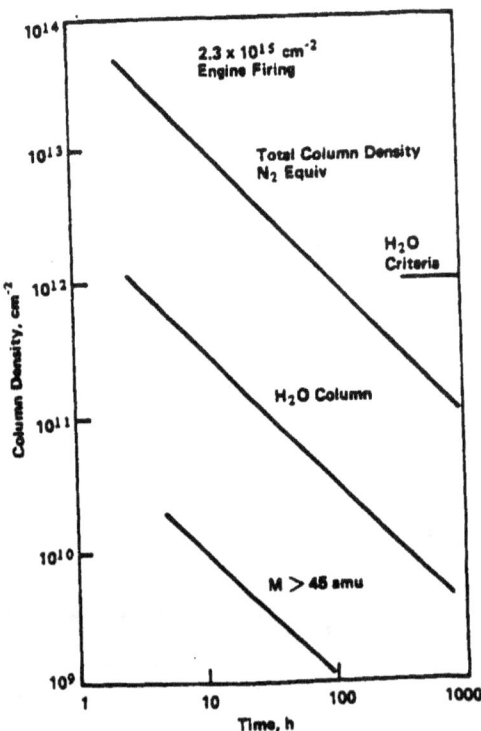

Figure 2.3-35 *Column Densities vs Time for Baseline STS Bay*

which is the ratio V/AC_D which produces a pressure differential when the external pressure drop rate is $\dot{P}$ and the pressure is $P+P_0$. In this expression $g = 9.81$ m/s², $R = 29.2$ m/K the gas constant for air and T is the absolute temperature. The coefficient C_D can be taken conservatively to be about 0.6. The evaluation of the required vent area, A, requires a knowledge of $\dot{P}$ and P_0 and the specification of an acceptable ΔP. The above equation provides the size of the vent area which allows a certain pressure differential to occur in a volume during the launch phase. Unless this area can be changed during other phases of flight and on the ground, the area size so established becomes basic to the purging flow requirements and the protection against contamination in the system.

The contamination prevention against external or internal contaminants to an instrument is predicated on the internal pressure and the exit velocity of the purging gas. The quantity of purge gas is a function of the vent area (A) and the pressure to be maintained (P), upstream of the vent area, and the downstream pressure P_0. For air or nitrogen, when $P_0 > 0.53P$ the viscous flow of gas at standard temperature and pressure (STP) from an orifice with small pressure differentials is given by

$$Q = \omega A v = \omega A [2gRT(P-P_0)/P_0]^{1/2} \text{ m}^3/\text{s} \qquad 2.3\text{-}11$$

which is obtained from the Torricelli equation. The coefficient ω may vary from 0.64 to 0.98 for an orifice. The flow velocity at the exit for $P_0 > 0.53P$ is

$$v = \alpha[2gRT(P-P_0)/P_0]^{1/2} = 24\alpha[T(P-P_0/P)]^{1/2} \text{ m/s} \qquad 2.3\text{-}12$$

The velocity coefficient α is about 0.98 for an orifice.

The continuum gas flow rate when $P_0 < 0.53P$ is

$$Q = 4.34 C_D \, AP/(T_0)^{1/2} \text{ m}^3/\text{s} \qquad 2.3\text{-}13$$

for A(cm²), P(torr) and T(°K). For T = 293°K this reduces to

$$Q = 0.253 C_D AP \text{ (m}^3/\text{s)} = 15.22 C_D AP \text{ (m}^3/\text{min)}. \qquad 2.3\text{-}14$$

This equation normalized with the volumes is depicted in Figure 2.3-36. The normalization provides the number of volume changes per unit time, Q/V (m³/m³/min) or the time needed for one complete volume change t_p = V/Q (min) as a function of the purging pressures and sizes of orifice. Molecular flow conditions exist when the gas mean free path is about ten times the diameter of the orifice. The exit flow velocity for the continuum flow regime when $P_0 < 0.53P$ is sonic at the orifice and is given by

$$v = \alpha[2g\gamma RT/(\gamma+1)]^{1/2} = 18.3\alpha(T)^{1/2} \text{ m/s} \qquad 2.3\text{-}15$$

where $\gamma = C_p/C_v = 1.4$ is the ratio of specific heats for air and $\alpha = 0.98$.

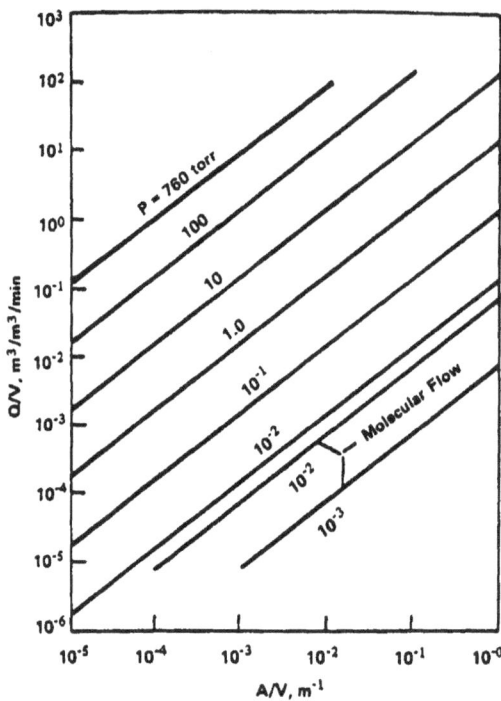

Figure 2.3-36
Volume Air Changes Q/V (min^{-1}) versus Vent Area Volume Ratio A/V (m^{-1}) as a Function of Internal Pressure

During the launch phase using a pressure controller, the purging pressure should be maintained at about 1 torr higher than the decaying external pressure. This will provide a flow out of the volume which will provide an obstacle to external contaminants. The flow will be 0.14 m^3/s with a velocity of 14.6 m/s at the start of the launch phase as calculated by Equations 2.3-14 and 2.3-15.

In orbit the purging pressure will be 1 torr and the bay pressure will be less than 0.53 torr. The in-orbit purge pressure must provide a sufficiently high pressure to prevent incoming gases and arrest, to a large extent, the internal outgassing. Under these conditions the purge flow will be 4.2×10^{-3} m^3/s and its exit velocity 309 m/s. This flow rate is equivalent to $Q = 3.18 \times 10^3$/torr sec being emitted in the bay.

In addition to the above purging techniques, heaters (to prevent deposition), cold traps (collecting molecules before they reach sensitive surfaces), materials selection, covers, and mission planning are all techniques which can be used for protection from orbital molecular contamination.

2.3.3.5 Key Technical Personnel

	Organization	Phone
L. J. Leger	JSC	(713) 483-2059
H. K. F. Ehlers (Modeling)	JSC	(713) 483-5539
H. W. Parker (Humidity Monitor and Dewpoint Hygrometer)	MSFC	(205) 453-0942
P. N. Peters (Air Sampler)	MSFC	(205) 544-7728
R. C. Linton (OEM and PSA)	MSFC	(205) 544-7725
J. Triolo (CMP)	GSFC	(301) 344-8651
D. McKeown (QCMs)	Faraday Labs	(619) 459-2412
E. R. Miller (IECM)	MSFC	(205) 544-7752
G. R. Carignan (Mass Spectrometer)	Univ. Michigan	
U. Von Zahn (SPAS)	Univ. Bonn	49-228-733235
L. E. Bareiss (Models/Analyses)	Martin Marietta	(303) 977-8713

2.3.3.6 In-Flight Molecular Contamination References

F-1. Peters, P. N., H. B. Hester, W. Bertsch, H. Mayfield, and D. Zatko, "Induced Environment Contamination Monitor (IECM), Air Sampler: Results from Space Transportation System (STS-2) Flight," SPIE Proceedings, Vol. 338, May 1982.

F-2. Miller, E. R., "Update of Induced Environment Contamination Monitor Results" paper 83-2582-CP, AIAA Shuttle Environment and Operations Meeting, Washington, D.C. Nov. 2, 1983.

F-3. Peters, P. N., H. B. Hester, W. Bertsch, and T. Marr, "Air Sampler", in NASA TM-86461, MSFC, August 1984.

F-4. Fountain, J. A., "Temperature-Controlled Quartz Crystal Microbalance and Cryogenic Quartz Crystal Microbalance", in NASA TM-82524, MSFC, February 1983.

F-5. Linton, R. C. and D. R. Wilkes, "Optical Effects Module and Passive Sample Array", in NASA TM-82524, MSFC, February 1983.

F-6. Linton, R. C., "Optical Effects Module and Passive Sample Array", in NASA TM-86461, MSFC, August 1984.

F-7. Fountain, J. A. and D. McKeown, "Quartz Crystal Microbalance Surface Contamination Measurements", in NASA TM-86461, MSFC, August 1984.

F-8. Triolo, J., R. Kruger, R. McIntosh, C. Maag, and P. A. Porzio, "Results from a Small Box Real-Time Molecular Contamination Monitor on STS-3", J. Spacecraft. and Rockets, 21, 400, 1984.

F-9. Miller, E. R. and G. R. Carignan, "Mass Spectrometer", SPIE Proceedings Vol. 338, May 1982.

F-10. Narcisi, R., E. Trzcinski, G. Federico, L. Wlodyka, and D. Delorey, "The Gaseous and Plasma Environment Around Space Shuttle", AIAA Shuttle Environment and Operations Meeting, AIAA-83-2659, Oct. 31, 1983.

F-11. Ehlers, H. K. F., S. Jacobs, and L. J. Leger, "Space Shuttle Contamination Measurements from STS-1 through STS-4", AIAA 21st Aerospace Sciences Meeting, Reno, AIAA-83-0331, Jan 10, 1983.

F-12. Carignan, G. R. and E. R. Miller, "Mass Spectrometer", in NASA TM-82524, MSFC, February 1983.

F-13. Carignan, G. R. and E. R. Miller, "Mass Spectrometer", in NASA TM-86461, MSFC, August 1984.

F-14. Wulf, E. and U. von Zahn, "Behavior of Contaminant Gases Emitted by the Space Shuttle and the Manned Maneuvering Unit", Minutes of Surface Interactions Subpanel, Space Shuttle Experiment and Environment Workshop, Aug. 8, 1984.

F-15. Parker, H. W., "Humidity Monitor and Dew Point Hygrometer", in NASA TM-82524, MSFC, February 1983.

F-16. "Skylab Contamination Prediction Report", ED-2002-1567, Rev. A, Technical Report, Martin Marietta, April 30, 1973.

F-17. Scialdone, J. J., "Correlation of Self-Contamination Experiments in Orbit and Scattering Return Flux Calculations", NASA TN D-8438, March 1977.

F-18. Ehlers, H. K. F., "Modeling Correlation with Flight Data", Shuttle Environment Workshop, Oct. 5-7, 1982.

F-19. Huang, S., "Flight Contamination Data Correlation Analysis", MCR-86-2002, Martin Marietta, January 1986.

F-20. Bareiss, L. E., F. J. Jarossy and J. C. Pizzicaroli, "Shuttle/Payload Contamination Evaluation Program: The SPACE Computer Program User's Manual/Final Report", MCR-80-593, Martin Marietta, September 1980.

F-21. Jensen C. L. and R. G. Goble, "Thermal Radiation Analysis System (TRASYS II)", MCR-73-105, Rev. 2, Martin Marietta, June 1979.

F-22. Scialdone, J. J., "Shuttle Measured Contaminant Environment and Modeling for Payloads", NASA TM-85111, Dec. 1983.

F-23. Simpson, J. P. and F. C. Witteborn, "Effect of the Shuttle Contaminant Environment on a Sensitive Infrared Telescope", Applied Optics, Vol. 16, No. 8, August 1977.

F-24. Scialdone, J. J., "Abatement of Gaseous and Particulate Contamination in a Space Instrument Application to a Solar Telescope", NASA TM-85016, April 1983.

Section 3—Surface Interactions

SECTION 3

SURFACE INTERACTIONS

Features of the Space Shuttle unique from earlier spacecraft designs have recently brought to light two nearly unanticipated physical phenomena occurring at Shuttle altitudes: 1) erosion of materials due to reactions with atomic oxygen (the dominant species at Shuttle altitudes); and 2) induction of a glowing region near the surfaces of some Shuttle materials exposed to the ram environment. Related to the surface glow is a prompt enhancement of glow associated with Orbiter thruster firings. The structure of the Orbiter PLB, exposing materials to the space environment, but protecting them during reentry, has allowed unprecedented observation of space environmental effects, including oxidation. The design of the PLB to allow complete astronaut observation has allowed greater viewing of surfaces exposed to the ram environment than on most previous spacecraft. This, along with the low typical Shuttle orbital altitude have resulted in greater awareness of the glow environment.

The following sections provide information on the current data and understanding of oxidation and glow phenomena.

3.1 MATERIALS OXIDATION/EROSION

3.1.1 Introduction

With the first Shuttle flight it became obvious that various materials exposed to the environment had undergone changes. Qualitative observations showed a loss of surface gloss (in particular Kapton), and an apparent "aging" of painted surfaces, as well as film thickness degradation; i.e., loss of material (Ref. G-1). To account for the latter, Leger suggested that atomic oxygen was reacting with materials (Ref. G-2).

Atomic oxygen is the dominant atmospheric species at STS and Spacelab altitudes. The orbital velocity of low earth orbit spacecraft (8 km/s) corresponds to a collisional energy of 5 eV for oxygen atoms. The flux of atoms impacting on a surface is obtained by multiplying the number density along the orbital path by the orbital velocity. Direct measures of atomic oxygen flux or density have not yet been made in conjunction with atomic oxygen effects flight experiments, so number densities for flux calculations have been obtained from atmospheric models for the time period of the experiments. The time integral of oxygen atom flux is termed fluence. Leger developed a quantitative factor for reaction characterization, the reaction efficiency (R.E.) which is derived by normalizing the thickness loss induced by the calculated oxygen atom fluence to yield R.E. = x cm^3/oxygen atom.

Two major experiments, termed Evaluation of Oxygen Interactions with Materials (EOIM) I and II were flown on STS-5 and STS-8. Additional data was collected on STS-3, STS-4 and STS-41G. The most sophisticated experiment yet developed to evaluate this phenomena (EOIM-III) is currently under construction for a future STS flight.

3.1.2 Available Oxidation/Erosion Data

Table 3.1-1 shows the R.E. results obtained for materials flown on STS-3, -4, and -5 (Ref. G-3), and Table 3.1-2 shows the results obtained in STS-8 (Ref. G-4). Atomic oxygen fluences for the three early flights are also shown in Table 3.1-1. The STS-8 fluence was 3.5×10^{20} atoms/cm^2. The similarity of the results is of particular interest. Commercial material names are shown, Kapton is polyimide, Mylar is polyethylene terephthalate, Tedlar is a copolymer of ethylene and tetraflouroethylene (TFE), and Teflon is either TFE or flourinated ethylene propylene (FEP).

Table 3.1-1 Atomic Oxygen Reaction Efficiencies (Ref G-3)

Shuttle Flight	Material	Thickness, μm (a)	Thickness Loss, μm	Fluence 10^{20} Atoms/cm^2	Reaction Efficiency 10^{-24} cm^3/Atom (b)
STS-3	Kapton TV Blanket	12.7	4.4	2.16	2.0
	Kapton, OSS-1 Blanket	25.4	5.5		2.5
STS-4 Witness Samples	Kapton MLI Blanket			0.65	
	Kapton	7.6	1.8		2.8
	Kapton	12.7	1.6		2.7
	Kapton	25.4	1.7		2.6
	Mylar	12.7	1.8		2.8
	Teflon FEP 7 TFE Al/Teflon FEP	12.7	0.07		0.1
STS-5 Witness Samples	Kapton	12.7	1.50	1.0	1.5
	Kapton	25.4	2.18		2.2
	Kapton	50.8	2.79		2.8
	Kapton, Black	25.4	1.35		1.4
	Mylar	12.7	2.16		2.2
	Mylar	25.1	1.83		1.8
	Mylar	50.8	1.50		1.5
	Tedlar, Clear	12.7	1.30		1.3
	Tedlar, White	25.4	0.41		0.4
	Teflon FEP & TFE	12.7	0.2		0.2
	Kapton (Coated)				
	DC1-2755	12.7 (Kapton)	0.2		0.2
	T-650	12.7 (Kapton)	0.2		0.2

(a) Note: Film Thicknesses of 12.7, 25.4, and 50.8 μm correspond to 0.5, 1.0 and 2.0 mils, respectively.
(b) Most probable error is +30 to 40%.

The dependence of reaction efficiency on temperature is not currently clear. No temperature dependence of reaction efficiency was found in the range 25°C to 125°C for the experiment reported in reference G-4. The experiment of reference G-4A, however, did find a dependence on temperature.

The qualitative observation of the "aging" of paints detected on STS-1 through -4 were extended on later flights with measurements of quantitative optical changes. The changes in emissivity ($\Delta\epsilon$) and absorptance ($\Delta\alpha_s$) were measured post-flight and are listed in Table 3.1-3 (Ref. G-3, G-5, G-6 and G-7). It can be noted that the oxygen inhibitors (silicone overcoat, labeled OI) and UV inhibiters (Irganox and Tinuvin) appear to have doubtful beneficial effects, and that silicone paints are hardly affected (Ref. G-5). The results for Chemglaze Z306 are of particular interest. They show a 4.8 percent weight loss whereas Auger spectroscopy showed a 400-500 percent increase in oxygen content (Ref. G-6). Of the last three entries in the table (conducting paints) only Electrodag 402 appears to be resistant to the environment at Shuttle altitudes. Note that it, too, contains a silicone.

Table 3.1-2 Recession and Reaction Efficiency for Organic Films on STS-8 (Ref G-4)

Material	Thickness, μm (MILS)	Exposed Side[a]	Surface Recession,[b] μm		Disc Samples	Average[c]	Reaction Efficiency 10^{-24} cm^3/atom
			Strip Samples				
			121°C	65°C			
Kapton	12.7 (0.5)	Air Roll	9.5 11.8	10.5 10.3	11.1		
Kapton	25.4 (1.0)	Air Roll	9.8 9.9	10.7 9.0		10.5	3.0
Kapton	50.8 (2.0)	Air Roll	11.1 11.1	10.6 11.1			
Mylar A	12.7 (0.5)	Air	12.7	12.3	12.7	12.6	3.6
Mylar A	40.6 (1.6)	Air	12.1	11.9		12.0	3.4
Mylar D	50.8 (2.0)	Air Roll	9.9 11.0	10.2 10.4		10.4	3.0
Clear Tedlar	12.7 (0.5)	Air	10.9	11.5		11.2	3.2
Polyethylene	20.3 (0.8)	N/A			11.5	11.5	3.3
Teflon TFE	12.7 (0.5)	Air			<0.2	<0.2	<0.05
Kapton F	30.5 (1.2)	N/A	<0.2	<0.2	<0.2	<0.2	<0.05

[a]Refers to Manufacturing Process: "Roll" Side in Contact with Manufacturing Rolls, "Air" Opposite Side.
[b]Corrected for Flux Reduction Due to Non Normal Impingement (cos θ)
[c]Strip Samples and Disc Samples

Table 3.1-3 Oxidation/Erosion "Aging" of Paints on STS-8

Paint	$\Delta\epsilon$	$\Delta\alpha$	Other Comments	Refs
A-276 Urethane, White	+0.03	−0.0007		
A-276 + 5% Ir (Ir = Irganox)	+0.02	+0.0007		
A-276 + 5% Ir + 2.5% Ti292 (Ti = Tinuvin)	+0.02	+0.016		
A-2767 + 5% Ir + 2.5% Ti900	+0.02	−0.006		G-5
V-200 Urethane	+0.02	+0.02		
V-200 + 5% Ir + 2.5% Ti292 + 2.55% Ti900	+0.02	+0.097		
V-200 + 2.5% Ir + 5% Ti292	+0.02	+0.057		
RTV-615 Silicone + TiO$_2$	−0.01	+0.0001		
RTV-615 + Carbon Black	0	0	Resistance Increase x2 per Unit Area	
Urethane + Carbon Black	+0.06	+0.0053	Resistance Increase x3 per Unit Area	
Flame Master S1023	−0.02	−0.02	11.3% Wt Loss; Oxygen Increase 25-50%	G-6
Chemglaze Z306	−0.02	+0.034	4.8% Wt Loss; Oxygen Increase 400-500%	
401-C10 (Black)		+0.005	Wgt Loss mg/O Atom 0.86 x 10^{-21}	
Z-853 (Yellow)		−0.034	0.9 x 10^{-21}	
GSFC (Green)		−0.002	No Change	
Z306 (Black)		+0.028	1 x 10^{-21}	G-7
Z302 (Glossy Black)		+0.043	5.8 x 10^{-21}	
Z302 + OI 650 Overcoat		−0.001	No Change	
Z302 + RTV 670 Overcoat		−0.004	No Change	
A276		−0.002	1 x 10^{-21}	
A276 + OI 650 Overcoat		+0.002	0.1 x 10^{-21}	
Electrodag 402 (Ag/Silicone)			2% Wt Loss	
Electrodag 106 (Gr/Epoxy)			68% Wt Loss	G-3
Aquadag E (Gr/Binder)			100% Wt Loss	

Table 3.1-4 is a compilation of results on various inorganic materials (Ref. G-8, G-9, G-10, G-11 and G-12). Up refers to out of the PLB, while down faces into the bay. The silicones again show high resistance, as noted above, to oxidative reactions, and the epoxy composites show R.E.s expected from organic materials (Tables 3.1-1 and 3.1-2). Additional realtime data collected during STS-4 and STS-8 with osmium and carbon coated QCMs are discussed in Section 3.1.5.1.

Table 3.1-4 Oxidation/Erosion of Various Materials

Flight	Material	Ref
STS-5	ZnS: "Up" Oxidized More Than "Down"; Deficient in Sulfur	G-8
	ThF_4: "Up" Shows Oxidation; ThO_2 Replaces ThF_4, Remaining ThF_4 is Stoichiometric	
	SiO: No Change	
	ITO (InSn Oxide: No Change	
	In_2O_3 (Coated Second Surface Mirror): No Change	
	Siloxane Coated KRS-5: Small Wgt. Loss, Roughened Surface	
	GFU-8 (Urethane Compound: 2.21% Wt Loss " Up", 3 Times More Than "Down"	G-9
STS-8	ZnS: Becomes $ZnSO_4$, No Change in IR Transmission	G-10
	ThF_4: Becomes ThO_2, No Change in IR Transmission	
	Silicone Grease: Visibly Intact but IR Shows Oxidation	
	Apiezon Grease: Essentially Gone by IR and Inspection	
	$AlMgF_2$: No Change in Reflectivity over 1216Å to 2200Å	G-11
	T300/5280 Epoxy Composite: IR Shows Oxidation, Reaction Efficiency = 2.9×10^{-24} cm^3/O Atom	G-12
	T300/934 Epoxy Composite: IR Shows Oxidation, Reaction Efficiency = 2.5×10^{-24} cm^3/O Atom	
	Gr/Al Metal-Matrix Composite: No Wgt. Change	
	Gr/Mg Metal-Matrix Composite: Wgt. Increase 0.40%	

On STS-5, evaluations were conducted on Kapton and Mylar films of various thicknesses in order to ascertain if thickness was an important parameter. Figure 3.1-1 shows that for Kapton the oxidation/erosion reaction increases with thickness, while Mylar shows the opposite effect. The apparent differences with thickness may be due to surface density variations associated with the manufacturing process (Ref. G-13).

An experiment on STS-8 was directed toward the mechanism of erosion. Electrically biased grids near the surface of an osmium film showed the erosion to be unaffected by the imposed bias voltage. It was thus concluded that the erosion was caused by atoms and not ions (Ref. G-4).

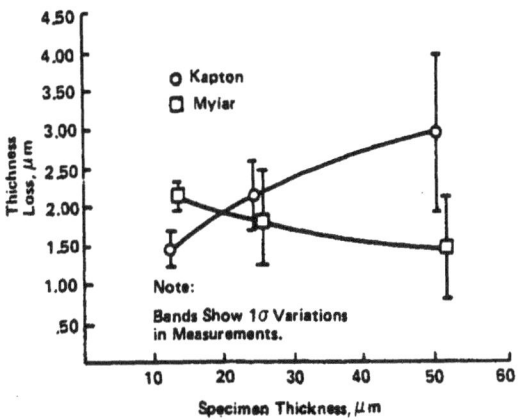

Figure 3.1-1 Erosion Versus Specimen Film Thickness, STS-5 (Ref G-13)

An investigation of the impingement angle of oxygen atoms on the oxidation/erosion of organic films was carried out on STS-8. Eleven specimens of Kapton (thickness 12.7 and 50.8 μm) and Mylar (thickness 12.7 and 40.6 μm) were inclined and positioned 42 degrees off the velocity vector axis and were exposed to only 74 percent of the normal impingement flux. The ratio of the combined recession data to that of normal impingement was 0.65 ± 0.03 rather than the expected 0.74 if the recession were due to a simple reduction in flux. Figure 3.1-2 shows the data fit to $(\cos \theta)^{1.5}$ is a better function than $\cos \theta$ (Ref. G-4).

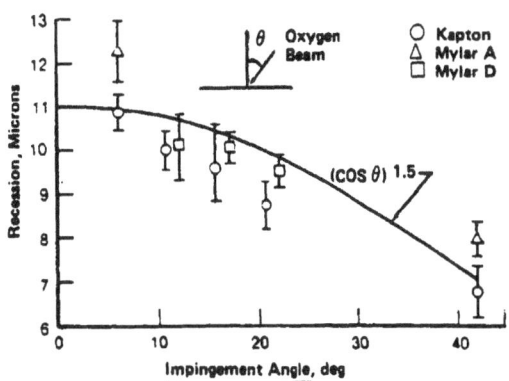

Figure 3.1-2
Surface Recession (Corrected for Flux Reduction) as a Function of Impingement Angle (Ref G-4)

The tensile strength of Kevlar 29 rope, a potential tether material for the Tethered Satellite, was examined on STS-8 and the results are shown in Table 3.1-5. As before, a coating of silicone appears to protect against deleterious effects of the oxidation/erosion environment (Ref. G-11).

Table 3.1-5
Oxidation/Erosion of Tensile Strength, STS-8 (Ref G-11)

Material	Tensile (lbs) After T/V Bake + Reabs H_2O*	After Flight + Reabs H_2O
Kevlar 29 Bare	696 ± 17	590 ± 15
Kevlar 29 Coated with DC1-2577	700 ± 20	677 ± 24
— Kevlar 29 with Jackets Removed	667 ± 24	671 ± 15

*T/V = Thermal Vacuum, Reabs = Reabsorption

A variety of elemental materials have been flown on Shuttle and the effect of the oxidation/erosion environment on various properties were investigated. The results are summarized in Table 3.1-6 (Ref. G-2, G-8, G-9, G-10, G-11, G-14, G-15, G-16, G-17 and G-18). Differences in observations with progressive flights are most likely due to increased sophistication used in the analysis of the results. For those materials (Ni, Ag) that showed a thicker oxide in the "down" compared with the "up" direction.

The general observations have been summarized as (Ref. G-19):

"1. Materials containing carbon (C), hydrogen (H), oxygen (O), and nitrogen (N) have high reaction rates in the range of 2.5 x 10^{-24} to 3.0 x 10^{-24} cm^3/atom.

2. Perfluorinated and silicone polymers are more stable than the organics by at least a factor of 50.

3. The reaction rates for filled organic materials are dependent on the oxidative stability of the fillers. For example, materials filled with metal oxides have lower reaction rates than those filled with carbon.

4. From a macroscopic standpoint, metals, except for osmium and silver, are stable. Metals such as copper do form oxide layers, but at lower rates than for osmium and silver."

Table 3.1-6 Oxidation/Erosion of Elemental Materials

Flight	C	Os	Ir	Al	Ni	Ag	Au	Pt, Pb, Mg Mo, W	Cr	Ref
STS-4	100-300Å Completely Removed R.E.=(0.14 - 0.43)	100-300Å Completely Removed R.E.=(0.14 - 0.43) $10^{1.25}$				2200Å Completely Oxidized	No Effect			G-2 G-14
STS-5		Auger Showed Some Os Remaining; Reflectivity Decrease by 300%	Auger Showed Ir and O; 10% Decrease in vac; U.V. Reflectance							G-8
				Small Increase in Directional-Hemisphere Scatter	Significant Decrease in Bi-Directional Reflectance Oxidation	Oxidation				G-9
STS-8				Anodized—No Change in Optical Properties						G-15
	R.E.=0.9 × $10^{2.5}$				Negligible Wt Change	Oxidized		Negligible Wt Change		G-11
						Oxidized and Flaked				G-16
		Oxidation Not Affected by Bias Grids		Oxidation		Oxidation and Sputtering	Sputtering			G-10
							Reflectance-Decrease 600-2000Å	Pt Reflectance Decrease 600-1200Å Increase 1200-2000Å	Oxidation	G-17
	2500Å Completely Removed	300Å Completely Removed								G-18

3-7

The Solar Maximum Recovery Mission (SMRM) has provided additional new data from long term space exposure. Fifty months in space at 310 nautical miles orbit has produced different effects on Kapton depending on its location in the experiments (Ref. G-20). No estimate of the oxygen atom fluence for the Solar Maximum exposure has been reported. The Power Supply Unit (PSU) fuse box cover lost 31.4 percent of its mass from its front surface and 3.51 percent from its rear surface. The Ground Support Equipment (GSE) test connector cover lost 7.4 percent and 0.54 percent from its front and rear surfaces, respectively. The lower weight loss for the rear cover in both instances has been attributed to shielding differently from direct impingment of reactive constituents.

Silver/Teflon films with Inconel overcoating the silver underwent drastic changes on the Solar Maximum satellite (Ref. G-20). Scanning electron microscopy (SEM) showed cracks in the Inconel and it was suggested that Ag_2O and Ag_2O_2 may have formed causing the degradation. Tensile tests of exposed samples showed the expected degradation of strength as a result of the loss of the metallic film. The results are compiled in Table 3.1-7 where the half/half specimen refers to a specimen cut 90° to the black specimen.

Table 3.1-7
Strengths of Eroded Ag/Teflon Film

Sample	Max Load	Elongation
1. Shiny (Least Eroded)	1.7 lb	125%
2. Half/Half	1.1 lb	50%
3. Black (Most Eroded)	1.1 lb	Zero

Tensile testing of unexposed silver/Teflon samples show that elongation is much greater than for the space-exposed film, even through the breaking loads are similar. Specimens were cut from an elongated piece of film, three in the "long" dimension and three more at 90° to that first direction, in order to emphasize any directionality. The average breaking load of the "long" dimension was 1.73 lb and the elongation was 210 percent; the breaking load for the samples in the 90° dimension averaged 1.75 lb and the elongation was 225 percent (Ref. G-20).

Samples of thick Teflon tape (5 mil) coated with 1500 Å of silver and 100 Å of Inconel were returned from SMRM and tested for tensile modulus (Ref. G-21). The preliminary results are given in Table 3.1-8.

Table 3.1-8
Tensile Modulus of Exposed Metalized Teflon

Sample	Exposure Conditions	Tensile Modulus	Appearance
1	High O atom, High UV	30% Decrease	Yellow
2	Medium O atom, No UV	15% Decrease	Slightly Yellow
3	High O atom, No UV	15% Decrease	Slightly Yellow

Teflon samples coated on the back side with Inconel protected silver under high magnification SEM showed near total silver reaction, flaking, and subsequent erosion of the underlying Teflon itself (Ref. G-22). A Kapton sample showed the pattern of a deliberate ethyl alcohol wipe presumably performed prior to launch.

While the mechanism of the Inconel/silver/Teflon film degradation has not been completely elucidated, many important features of the process have been identified. The transparent material has a cracked mosaic-like surface structure which can be attributed to thermal effects. Unexpectedly, only trace quantities of oxygen are observed on the surface. A reaction product composed primarily of silver, carbon, fluorine and chlorine is observed protruding from the cracks. This material is easily detached leaving the underlying regions exposed. During oxygen exposure, most of the Inconel layer and much of the silver layer are removed. In addition, Secondary Ion Mass Spectrometry (SIMS) and X-ray Photoelectron Spectroscopy (XPS) analyses indicate that a Teflon reaction has taken place. Fluorine is liberated resulting in a fluorine depleted Teflon surface. While this reaction is not fully understood, it is likely that silver in the presence of atomic oxygen catalyzed the breakdown of the Teflon. (Ref G-23).

Aluminized Kapton and silvered Teflon were tested on the SMRM where samples were exposed to atomic oxygen and UV simultaneously and to atomic oxygen only (Ref. G-24). The aluminized Kapton had been wiped with an alcohol based solvent prior to launch. For specimens exposed simultaneously to UV and atomic oxygen the surface (as shown by SEM) showed more damage in the "wiped" areas when compared to adjacent "non-wiped" areas. Microscopic tracks were also observed which on detailed examination showed sub-micron sized craters and holes in the tracks. Samples exposed only to atomic oxygen had surface morphologies similar to those exposed to atomic oxygen and a limited amount of UV. However, the latter had much larger features. The variation of atomic oxygen impingement angle and UV exposure are suggested as creating this difference.

For silvered Teflon exposed to atomic oxygen and UV, the Teflon side showed "cone" formations of Teflon degradation in direct contrast to previous exposure experiments on Shuttle which showed Teflon to be extremely stable. The specimens which were exposed in the velocity vector of atomic oxygen with no UV exposure had a surface roughness compared to unexposed Teflon but the damage is insignificant when compared to specimens exposed simultaneously to UV and atomic oxygen. It thus becomes important to understand synergistic effects of the entire space environment, including temperature, atomic oxygen, and radiation (low earth orbit, solar UV and particulate, and cosmic) (Ref. G-24).

Additional oxidation data of interest was collected on STS-41G (Ref. G-25). In addition to the protection data of Section 3.1.6, significant Kapton and composite material results were collected. In the case of Kapton, samples manufactured in circa 1969 were found to exhibit significantly higher resistance to oxidation than samples manufactured in circa 1984. Chemical analyses indicated only minor compositional differences in the samples. A more significant identified difference is a heat treating process for surface passivation and surface tension relief discontinued by the manufacturer in the late 1960s. How this or other differences could offer resistance to atomic oxygen warrants further investigation. Carbon-epoxy samples revealed a reaction efficiency of approximately 2×10^{-24} cm^3/atom for the epoxy matrix. Once exposed, the graphite fibers also showed high susceptibility to oxidation. STS-41G fluence was approximately 3×10^{20} atoms/cm^2.

3.1.3 Additional Oxidation/Erosion Data

3.1.3.1 Unpublished Oxidation/Erosion Data

No unpublished data is currently identified, although not all the samples and experiments flown on EOIM-I and EOIM-II have been reported in the open literature.

3.1.3.2 Needed Oxidation/Erosion Data

The recession rates (or reaction efficiencies, R.E.) to date have used models of the atmosphere (Section 4.0) for the fluence calculations. This can lead to errors ranging from ±25 percent to factors of 2 (Ref. G-26). Therefore, it is necessary to measure the ambient density in orbit concurrently with erosion effects in order to systematically examine the parameters affecting the mechanisms of erosion. A mass spectrometer system to look at the ram ambient environment and then to examine the species evolving from surfaces of various materials is required.

3.1.3.3 Planned Experiments

An instrument designed to collect the above required data is in the late planning stages and was to be flown in late 1986 or early 1987 (Ref. G-26). The extent of launch slippage due to the STS-51L accident is unknown at the time this Handbook edition is being prepared. An AFGL mass spectrometer will be used to examine the ram ambient environment and the environment near samples on exposure trays. The instrument is called Evaluation of Oxygen Interaction with Materials (EOIM) III. The instrument is shown in Figure 3.1-3 (Ref. G-27)

In addition, the current data base is quantitative at best. An ideal data base would include information on oxidation effects such as changes in solar absorptivity, emissivity, transmissivity, reflectivity (including specularity changes), resistivity, strength, and any other parameters that could be effected by chemical reaction with oxygen. An ideal data base would also include information on how variables such as flux level, incidence angle, and temperature effect reaction rates, as well as how other parameters such as radiation and vacuum exposure change the rates. Finally, the above information in an ideal data base would be included for the entire inventory of materials currently in use or proposed for use on spacecraft. The extent of the data needed implies that ground simulators of the space environment will be necessary to obtain a complete data base.

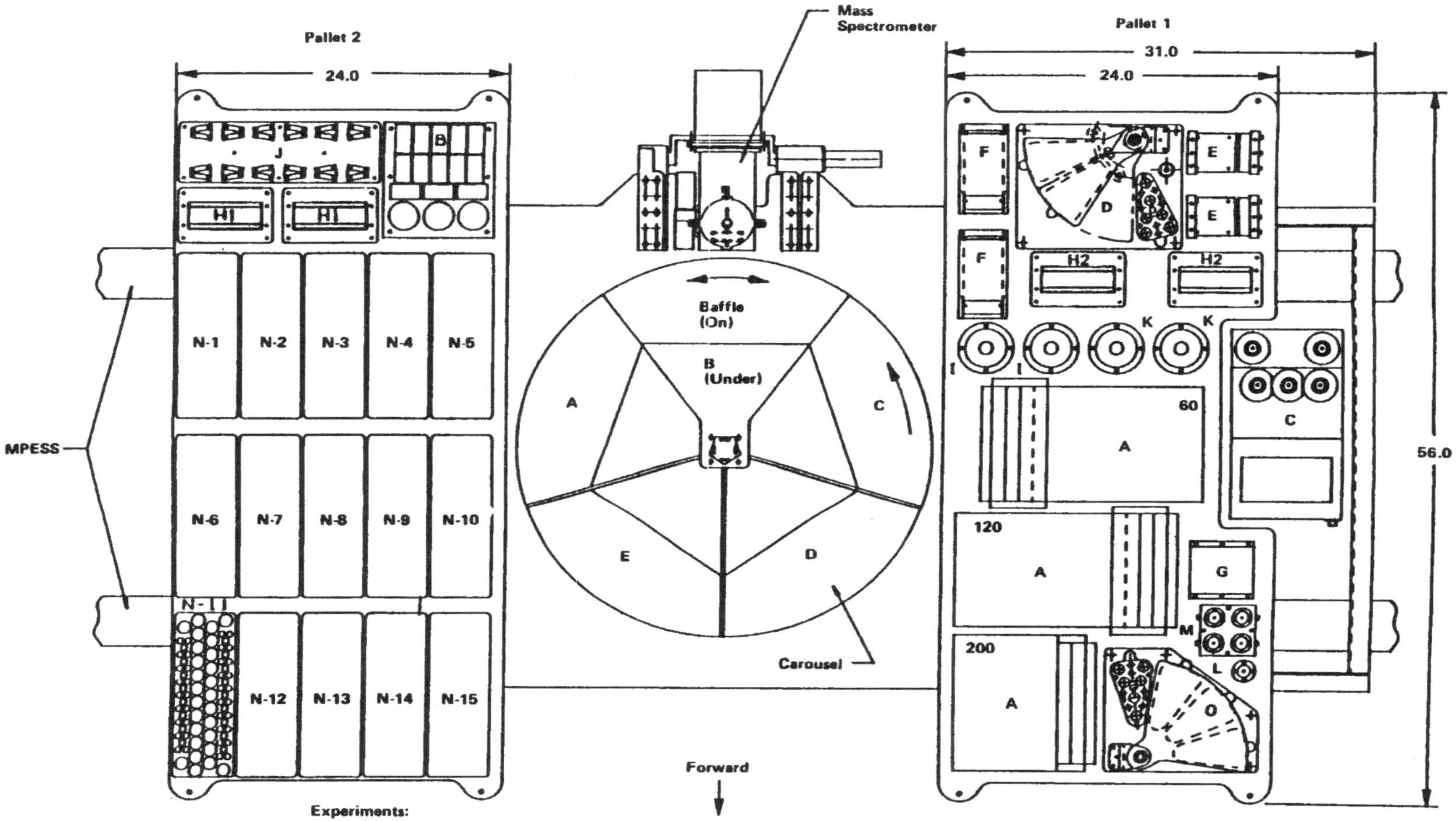

Experiments:

A — Heated Plate (JSC), 3 each
B — Atom Scattering Experiment (UAH), 1 each
C — Environment Monitor Package (GSFC), 1 each
D — Solar UV Experiment (JSC), 1 each
E — Static Stress Fixture (MSFC), 2 each
F — Uniform Stress Fixture (MSFC), 2 each
G — Atomic Oxygen Monitor (MSFC), 1 each
H1 — Composite Stress Fixutre (LaRC), 2 each
H2 — Composite Stress Fixture, (JSC), 2 each
I — Scatterometer (JPL), 2 each
J — Mechanical Stress Fixture, (LeRC), 11 each
K — Reflectometer (LeRC), 2 each
L — Pin-Hole Camera (LeRC), 1 each
M — Scatterometer (Aerospace Corp), 1 each
N — Sample Carriers, 15 each
O — Variable Exposure Tray, 1 each

Figure 3.1-3 Evaluation of Oxygen Interaction with Materials Experiment III

3.1.4 Ground Oxidation/Erosion Simulation

A laboratory study of the rate of reaction of solid carbon with atomic oxygen atoms, using an ion beam impacting at energies near 1 eV, has been measured at carbon surface temperatures of 300-400K (Ref. G-28). The targets were amorphous carbon films approximately 300Å thick deposited on fused silica disks. The reaction was monitored in situ by measuring the changes in absorptance, and hence, thickness of the carbon film as a function of O atom beam exposure time. Figure 3.1-4 shows an Arrhenius plot of the results. Data from two flight experiments (Ref. G-4 and G-13) and another laboratory experiment (Ref. G-29), are also included. The line labeled "Park" is the equation

$$P = 0.63 \, e^{-1160/T}$$

which Park used to describe the results of six different investigations of the oxidation of various graphitic materials for surface temperatures ranging from 300-4000K (Ref. G-30). The data of Reference G-28 produces the equation

$$P = 4.2 \, e^{-1800/T}$$

where P is defined as a reaction probability. This equation implies an activation energy of 15 kJ/mole (3.6 kcal/mole) in reasonably good agreement with Park's value of 10 kJ/mole (2.4 kcal/mole).

It should be noted that these activation energies are surprisingly small when compared to the usual activation energies for ordinary chemical reactions and/or desorption activation energies for various gases or metals (Ref. G-31).

The same experimenters have studied the rate of removal of Kapton with impacting oxygen atoms at 1 eV (Ref. G-32). The samples were disks of 5 mil thick Kapton, 0.9 inch in diameter. The experimental results demonstrate that the rate of removal of Kapton in LEO can be approximated in the laboratory. Table 3.1-9 presents a comparison of the laboratory results with the results from STS-5 and STS-8. The results suggest that there is not a great dependence on the oxygen atom translational energy and the reaction rate, although this conclusion requires much further investigation with this and other materials prior to acceptance (see also Ref. G-4).

Table 3.1-9
Comparison of Laboratory and Flight Measurements of the Average Probabilities for the Reaction of Atomic Oxygen with Kapton

Kapton Temperature, K	Reaction Efficiency 10^{-24} cm^3/O Atom		
	Ground Test (Ref G-32)	STS-5 (Ref G-5)	STS-8 (Ref G-16)
300	2.1 ± 1.1 1.7 ± 0.9	2.3 ± 0.9	—
338	1.4 ± 0.9	2.0 ± 0.8	3.0 ± 1.2
393	1.5 ± 0.9	2.1 ± 0.9	2.9 ± 1.2

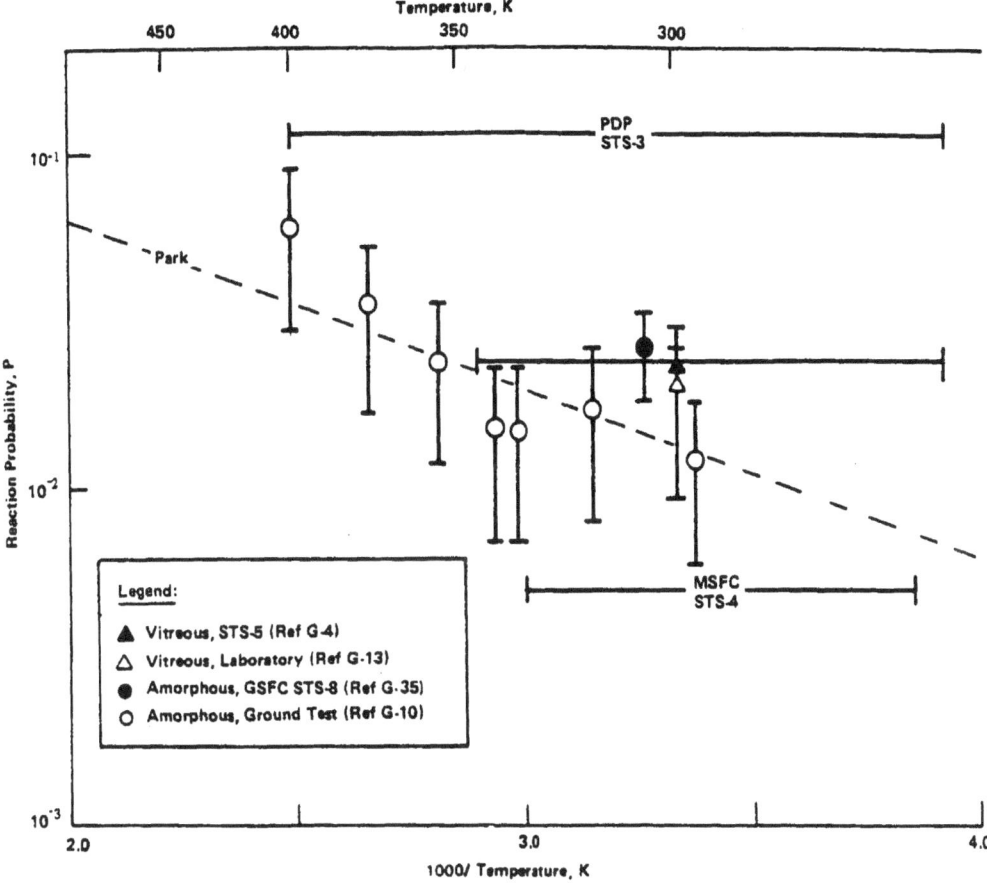

Figure 3.1-4 Arrhenius Plot of the Reaction Probability of the O + Carbon Reaction (Ref G-10)

The results of an early laboratory study of reactions of atomic oxygen with various polymers are shown in Table 3.1-10 (Ref. G-33). The study was carried out by placing polymer specimens on glass slides in the sample chambers of a Tracerlab low temperature asher Model 500A. The sample temperatures were $\leq 40°$. The oxygen atom concentration was on the order of 10^{14} to 10^{15} atoms/cm^3 at a pressure of 1 torr, and the flow of oxygen was about 4 cm^3 (STP)/min. As with flight experiments, polyimide is one of the least reactive materials listed in the table. The flourinated materials, which can be considered as the analog to the flight silicones, are the least reactive, verifying the inhibition of silicone coatings to atomic oxygen reaction as observed on Shuttle experiments.

Similar tests were conducted on a proposed Shuttle space suit material officially named ST11G041-01, Shell TMG two-layer plain weave "orthofabric", by exposing it to an asher discharge (Ref. G-34). The material is constructed of polymeric yarns of Nomex and Kevlar-29. The asher (SPI Plasma Prep II) provided atomic oxygen ions under a 13.6 MHz microwave discharge. Air pressure in the asher during tests was about 140 μm. There were no diagnostics available to measure either plasma density or ion temperatures, nor to know whether the density and temperature fluctuated during a test run. The fabric was exposed to this simulated environment for about 17 hrs and the mass loss was determined. The mass loss rates were then scaled to reflect the ionosphere at about 220 km. The results of the study predicted the mass loss in the ionosphere to be about 66 percent of the original fabric mass/year, assuming a flux of 7.44 x 10^{22}/cm^2 yr. Caution must be used in applying asher generated test data due to the facts that: 1) the 5 eV orbital atomic oxygen energy is not simulated (simulation energy is actually 1 to several hundred eV); 2) flux is not simulated; 3) the flux is contaminated with O_2, O_2^+, O^+, and possibly other species; and 4) the flux is not unidirectional. In addition, extrapolating flight predictions from asher results is valid only if the material being tested and the flight tested material used for scaling both have the same correspondence between flight and asher response. The only valid application of asher tests may be in direct comparisons of relative effects on materials. The applicability of asher tests to quantitative analysis or accurate design support in predicting materials lifetimes has not been demonstrated.

Table 3.1-10 Atomic Oxygen Reaction with Polymers (Ref G-32)

Type of Polymer	g x 10^4/4.84 cm^2/min
Low Density Polyethylene	2.48
Irradiated Low Density Polyethylene (1 Mrad)	2.77
Irradiated Low Density Polyethylene (10 Mrad)	3.41
Irradiated Low Density Polyethylene (105 Mrad)	4.12
Chemically Crosslinked Low Density Polyethylene	3.31
Low Molecular Weight Highly Branched Polyethylene	3.26
High Density Ethylene-Butene Copolymer	3.09
Polypropylene	3.45
Polybutene-1	3.56
Chlorinated High Density Polyethylene	5.00
Chlorinated Polyethylene Plus 10% Polysulfide Polymer	2.90
Natural Rubber	3.39
Natural Rubber-Sulfur Raw Stock	1.20
Natural Rubber-Sulfur Vulcanizate	0.16
Natural Rubber-Peroxide Raw Stock	2.99
Natural Rubber-Peroxide Cured	1.67
Commercial Hard Rubber	2.71
Vulcanized Ethylene-Propylene Rubber	0.20
Polystyrene	1.26
Poly-3-Phenyl-1-Propene	1.43
Poly-4-Phenyl-1-Butene	1.67
Polyvinylenecyclohexane	2.28
ABS Polymers, Several Types	2.68
Unplasticized Polyvinyl Chloride Copolymer	4.71
Polyvinyl Fluoride	2.54
Polytetrafluoroethylene	0.82
Perfluorinated Ethylene-Propylene Copolymer	0.44
Polymethyl Methacrylate	2.14
Polyimide	1.19
Polycarbonate	2.59
Polyethylene Terephthalate	1.82
Nylon G	2.77
Nylon 610	3.24
Formaldehyde Polymers	5.77-7.85
Polysulfide (Chloroethyl Formal Disulfide)	19.45
Cellulose Acetate	5.0

Several other groups, including JPL, Physical Sciences, Inc. (PSI), Martin Marietta, and Los Alamos National Laboratory (LANL), are preparing to undertake laboratory studies of oxygen impingement on Shuttle materials. Details of some of these experiments can be found in Ref. G-35. Additional work is ongoing at Martin Marietta to develop a neutral atomic oxygen beam facility for MSFC (Ref. G-36).

3.1.5 Oxidation/Erosion Models

3.1.5.1 Oxidation Mechanism Models

An insight into the reaction efficiency (R.E.) as defined by Leger (Ref. G-2) can be obtained by considering the oxidation/erosion of films of carbon and osmium that were flown on STS-4 and STS-8.

On STS-8, a coated TCQM was used to measure the quantitative oxidation/erosion of carbon and osmium films (Ref. G-18). For the carbon a 2500Å layer of carbon was deposited on the TQCM, and the osmium film was 300Å thick. Figure 3.1-5 shows a linear loss (i.e., constant with time) for the carbon film (Ref. G-18). Figure 3.1-6 shows the loss of the osmium film (Ref. G-18). In contrast with carbon, the osmium loss appears to take place in two steps, neither of which are linear with time. Data extracted from Figure 3.1-6 is presented in Figure 3.1-7 showing an exponential removal for both steps.

The kinetics of heterogeneous reactions (Ref. G-37) may be used to discuss such results. In a system consisting of a solid surface and a gas striking it, as a general rule, the latter will "condense" for a period of time. Then, as a result of thermal agitation, "evaporation" will take place from time to time. If α is the fraction of the gas which adheres, then $\alpha\mu$ is the number which "condense" on each cm^2 of available surface per second, where μ is the number striking one cm^2 per second. If θ is the fraction of the total surface covered with gas at any instant, then $1-\theta$ is the fraction of uncovered surface. Assuming that only a single layer of gas can form on the surface, the rate of condensation will be $(1-\theta)\alpha\mu$ per cm^2 per second. The rate of "evaporation" will be proportional to the area covered, θ, so that it may be represented by $\nu\theta$ where ν is a constant for the gas-surface system. When the rates of condensation and evaporation are equal:

$$(1-\theta)\alpha\mu = \nu\theta \qquad 3.1\text{-}1$$

$$\theta = \alpha\mu/(\alpha\mu + \nu). \qquad 3.1\text{-}2$$

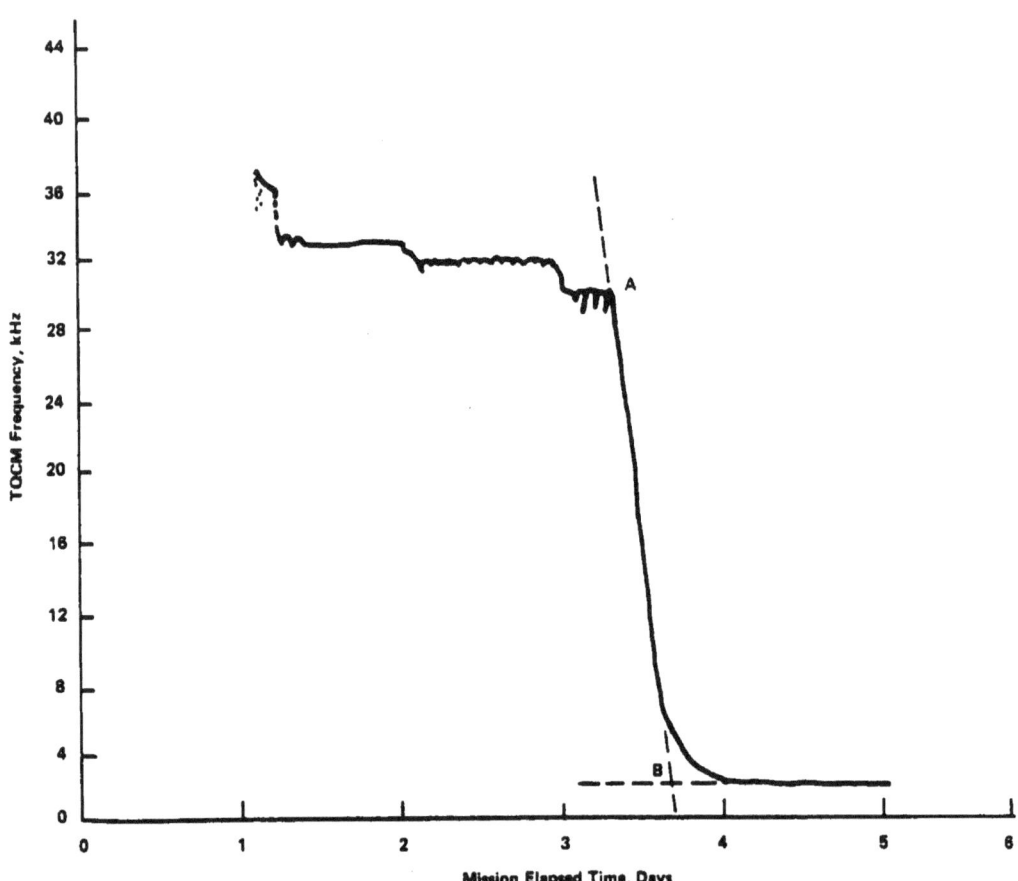

Figure 3.1-5
STS-8 Atomic Oxygen Monitor TQCM 2: Carbon Coated (Facing Out of Bay) (Ref G-18)

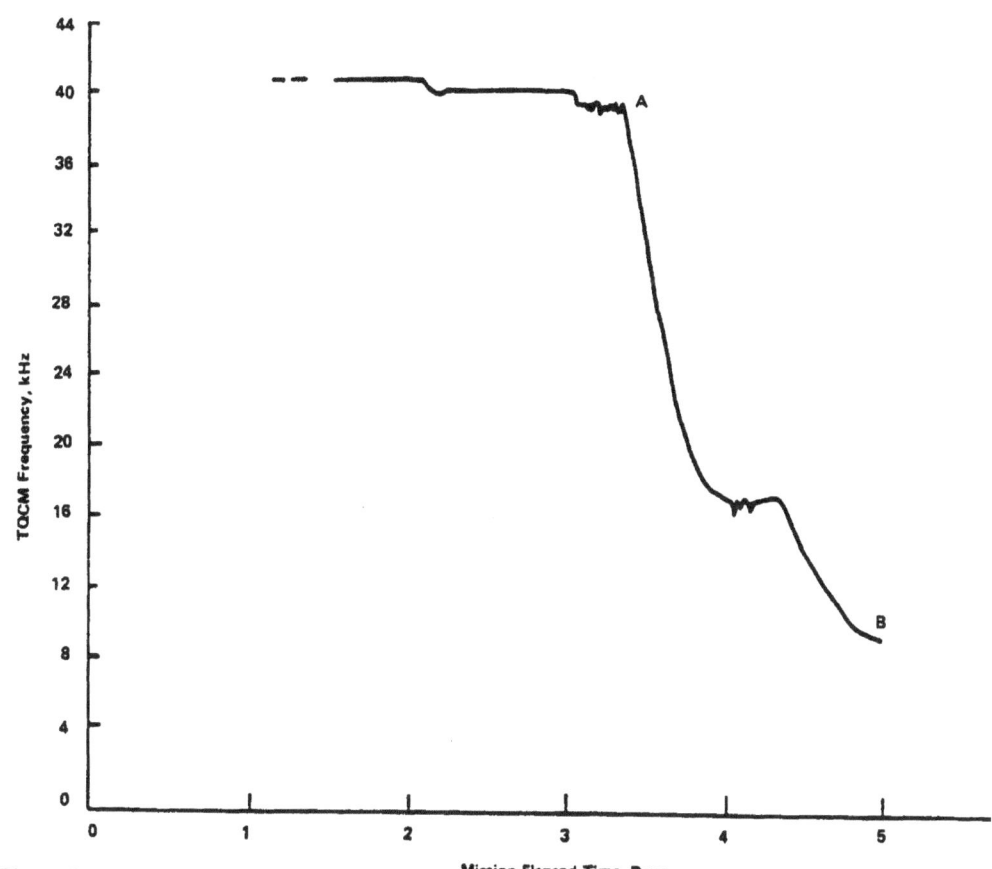

Figure 3.1-6
STS-8 Atomic Oxygen Monitor TQCM 4: Osmium Coated (Facing Out of Bay) (Ref G-18)

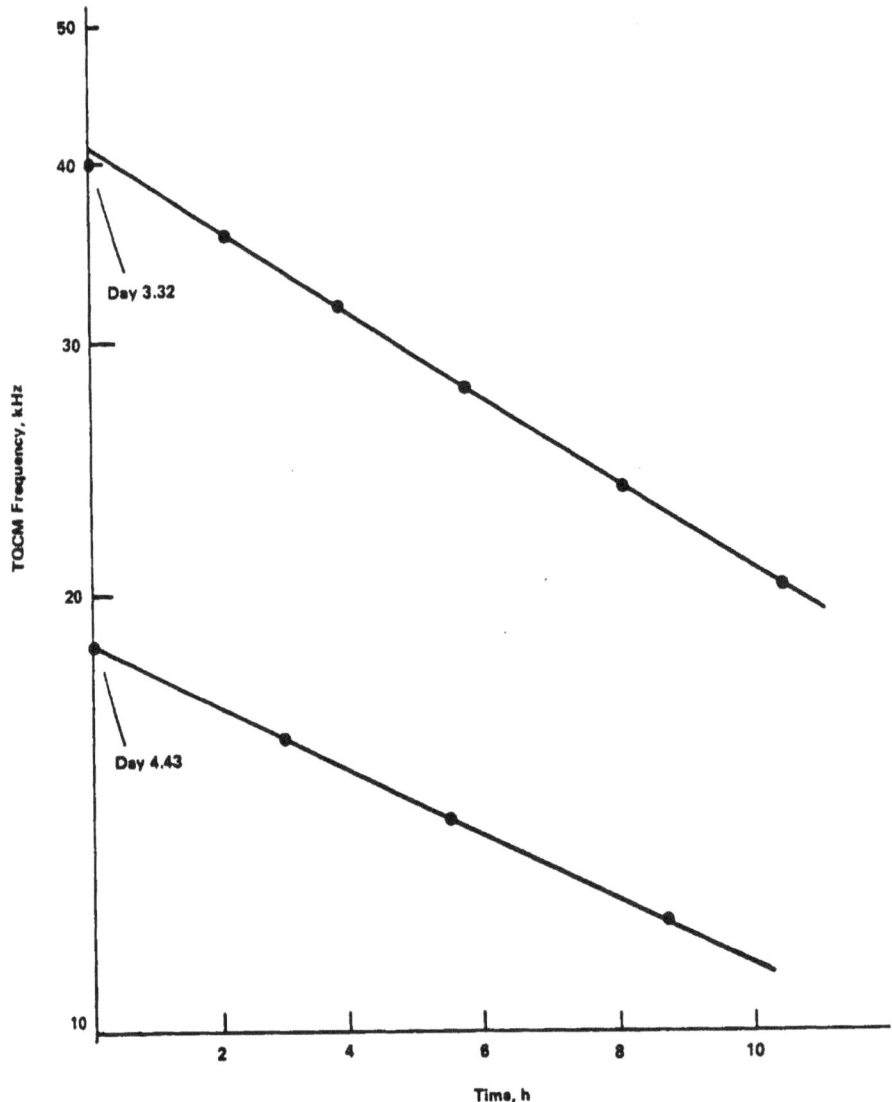

Figure 3.1-7 Kinetics Analysis of TQCM Data of Figure 3.1-6

In general, if reaction is followed immediately by evaporation of products, the rate of reaction is given by:

$$dx/dt = \nu_1 \theta^m \qquad \qquad 3.1-3$$

where ν_1 determines the rate of evaporation of the products and θ is a measure of the surface concentration of the reacting molecules. In the

3-18

simplest case, m (the power of θ) is equal to unity so that:

$$dx/dt = \nu_1\theta. \quad\quad 3.1-4$$

Substituting the value of θ from Equation 3.1-2:

$$dx/dt = k_1\mu/k_2\mu + 1 \quad\quad 3.1-5$$

where k_1 and k_2 are constants proportional to $\nu_1\alpha/\nu$ and α/ν, respectively. With the assumption made above that only one gas molecule (or atom) is involved in the reaction, the process is unimolecular, but examination of Equation 3.1-5 shows that it is not a simple "first order" (chemical kinetics terminology) or the 1st power in the concentration of one of the reacting components (i.e., m = 1 in Equation 3.1-3). When the surface is almost completely covered (i.e., θ-->1), then Equation 3.1-3 becomes dx/dt = constant (since ν_1 is a constant), the rate of reaction will be independent of concentration, and the process is kinetically of "zero order" (i.e., m = 0 in Equation 3.1-3). This appears to be the case for carbon loss in this approximation of heterogeneous kinetics.

If the surface is sparsely covered (i.e., α is small), but evaporation is rapid (i.e., ν is large), θ will be small in comparison with unity; then Equation 3.1-1 becomes $\alpha\mu = \nu\theta$ and combination with Equation 3.1-4 gives:

$$dx/dt = k'\mu = k. \quad\quad 3.1-6$$

Thus a unimolecular heterogeneous reaction becomes kinetically of first order. Chemical reactions of the first order behave with time as depicted in Figure 3.1-6, and it appears that the loss of osmium can be described in this manner for both steps of the reaction (Figure 3.1-6).

When the surface is almost completely covered (θ-->1), Equation 3.1-2 becomes $\alpha\mu = \alpha\mu+\nu$, and Equation 3.1-5 simplifies to:

$$dx/dt = k_1\mu/(k_2\mu+1) = (\nu_1\alpha\mu)/(\alpha\mu+\nu) = \nu_1$$

$$= \text{constant} = k \quad\quad 3.1-7$$

since ν_1, is constant, and the reaction is kinetically of "zero order" (linear in time). The rate constant (k) for atom loss has units of atoms/time.

Just as a unimolecular reaction becomes of zero order when one of the products is firmly held on the surface, so a bimolecular process (1st power in each of the components or 2nd power in one of them) may, for the same reason, prove to be kinetically of the first order. Without question the osmium loss is more complex than the carbon loss (compare Figures 3.1-5 and 3.1-6). The analysis of Figure 3.1-7 shows it to be kinetically of first order; this, however, does not preclude a bimolecular reaction. If the osmium is removed by chemical reaction with oxygen atoms and not simple sputtering (erosion) and leaves as OsO, the discussion indicates that evaporation is rapid (and the surface is sparsely covered with the product). If it leaves as OsO_2, then the product is rather firmly held, and the bimolecular process (two oxygen atoms) appears kinetically as first order.

Figure 3.1-5 shows the carbon loss in flight to be linear for over 90 percent of the film thickness, i.e. between points A and B. Thus, the oxidation is "zero order" kinetically. From the above simplifications of Equation 3.1-5 it would appear that the reaction is described by Equation 3.1-7. However, this requires the surface to be almost completely covered with the reacting molecules. Almost certainly the surface is sparsely covered, since the product of reaction is expected to be CO which has a very high vapor pressure. Sparsely covered surfaces are described by Equation 3.1-6 which is for a "first order" reaction in contrast to the observed zero order. In the space environment, however, the collision frequency of oxygen atoms with the surface is constant and the right-hand side of Equation 3.1-6, therefore, becomes constant. Thus, in space, the first order reaction with sparsely covered surfaces is reduced to a zero order because of the constancy of the collision frequency.

To account for thickness loss of materials, Leger defined the Reaction Efficiency (Ref. G-2). Implicit in the definition is the assumption of zero order kinetics since it defines the thickness loss as linear in time (since the fluence contains time) or Δ thickness/Δ time = constant. Clearly for films where the thickness loss is complete, such as for carbon and osmium, the Leger R.E. can be in error. For example, on STS-8 the fluence is determined as 3.5×10^{20} oxygen atoms/cm^2 from the 41 hours (1.5×10^5 secs) in the RAM direction but the carbon was lost in only 3×10^4 sec (between points A and B in Figure 3.1-5) and the osmium in 10.1×10^4 sec (between points A and B in Figure 3.1-6). The Leger R.E. for carbon is therefore calculated as 2500 Å film thickness loss/3.5×10^{20} = 0.71×10^{-25} cm^3/atom and for osmium it is 300 Å /3.5×10^{20} = $.086 \times 10^{-25}$ cm^3/atom. When corrected for the actual time, i.e. $1.5 \times 10^5/3 \times 10^4$, the carbon reaction efficiency becomes some 5 times larger, and for osmium it is 1.5 times larger.

It should be noted that even this is not totally accurate for osmium since the osmium kinetics are not zero order but rather first order for both steps in the loss (see Figure 3.1-7).

The rate constant for the zero order kinetics of carbon is easily obtained. The area of the 2500 Å thick sample was 5.06 cm^2 for a total volume of carbon 12.65×10^{-5} cm^3. From the density of graphite (2.267 g/cm^3) and the molar volume = 5.3 cm^3, the total number of carbon atoms lost is ($12.65 \times 10^{-5}/V_m$) 6.02×10^{23} = 14×10^{18} atoms. Thus, the rate constant, k, becomes 14×10^{18} atoms/3×10^4 sec = 4.4×10^{14} carbon atoms/sec. The number of oxygen atoms required to remove a carbon atom is given by: flux/rate constant = $2.32 \times 10^{15}/4.4 \times 10^{14}$ = 5.2.

Alternatively, a reaction probability, P, can be defined as P = R.E. x (area of loss)/volume of carbon atom = 0.207 and then $1/P = 4.8$, in good agreement with that obtained from the rate constant. This is, of course, essentially equal to: correct fluence/number of carbon atoms = 3.5×10^{20} ($3 \times 10^4/1.5 \times 10^5$) = $7 \times 10^{19}/1.4 \times 10^{19}$ = 5. The latter is the simplest method for determining an efficiency defined as the number of oxygen atoms required to remove one carbon atom.

For the osmium loss there appears to be two separate first order reactions (see Figure 3.1-6). The rate constants may be calculated from the half life, $t_{1/2}$, for each reaction. In the first step, half of the thickness (41kHz/2) is lost in 3.8×10^4 seconds so that the rate constant is $k = \ln 2/t_{1/2} = 1.8 \times 10^{-5}$/sec. For the second step, half of the remaining 62 Å (18.5kHz/2) is lost in 4.3×10^4 sec so that $k = 1.6 \times 10^{-5}$/sec. (The closeness of those values may indicate some malfunction of the quartz microbalance for a time, near day 4 (Figure 3.1-5), with subsequent recovery.)

The first order kinetic equation (Equation 3.1-6) can be written for the loss of osmium as

$$dx/dt = k(a_0 - x) \qquad 3.1\text{-}8$$

where a_0 is the original amount of osmium and x is the amount lost so that $a_0 - x$ is the amount remaining at time t. Equation 3.1-8 can be integrated to yield

$$a_0 - x = a_0 e^{(-kt)} \qquad 3.1\text{-}9$$

or $\quad (a_0 - x)/a_0 = e^{(-kt)}$ = fraction remaining $\qquad 3.1\text{-}10$

so that $\quad 1 - e^{(-kt)}$ = fraction lost at time t. $\qquad 3.1\text{-}11$

For example, to calculate the time to 99 percent loss for the first reaction (Figure 3.1-5)

$$1 - e^{(-1.8 \times 10^{-5} t)} = 0.99$$

$$-1.8 \times 10^{-5} t = -4.61$$

or $\quad t = 4.61/1.8 \times 10^{-5} = 2.56 \times 10^5$ sec.

For the second reaction the time to 99 percent loss of the remaining 62 Å of film is

$$-1.6 \times 10^{-5} t = -4.61$$

or $\quad t = 4.61/1.6 \times 10^{-5} = 2.88 \times 10^5$ sec.

If there were no malfunction of the quartz crystal microbalance during loss of the osmium, then the reaction efficiency could be instructive. For the first step (Figure 3.1-6), the correct fluence = (flux) x (time) = 2.33×10^{15})(2.56×10^5) = 6×10^{20} oxygen atoms/cm^2. From the film thickness loss of 238 Å and the volume of the osmium atom (r = 1.34 Å), the total number of osmium atoms lost is calculated as 8.7×10^{17}. Therefore > $6 \times 10^{20}/8.7 \times 10^{17}$ = 690 oxygen atoms are required to remove one osmium atom. In the second step 2.88×10^5 sec are required to remove the final 62 Å of film. The fluence is calculated as 6.7×10^{20} oxygen atoms/cm^2, and the total number of osmium atoms removed is 3×10^{17}, so $6.7 \times 10^{20}/3 \times 10^{17}$ = 2233 oxygen atoms are required to remove one osmium atom. From the ratio 2233/690 = 3.2 it may be assumed that three times as many oxygen atoms are required in the second step compared to the first step. If it is assumed that in the first step the osmium is lost as OsO, then the osmium loss in the second step probably involves OsO$_3$. The reason for the change in the chemistry (if real) is unclear.

Figure 3.1-8 shows laboratory results of the effects of atomic oxygen on various polymers (Ref. G-33). The figure shows the reactions to be linear in time (i.e., zero order). Thus, the implicit assumption of zero order kinetics in the Leger R.E. appears to be justified. Table 3.1-10 from the same study shows weight loss/cm^2 sec for a variety of polymers. In the study the oxygen atom concentrations were not quantitatively measured but were estimated to be in the range 10^{14} to 10^{15} atoms/cm^3, at a pressure of 1 torr flowing over the sample at 4 cm^3/min. Flowing of the gas ensures the collision frequency to remain essentially constant as in the space environment. From the table the values of polyimide and polyethylene terephthalate may be compared with the flight specimens Kapton and Mylar listed in Tables 3.1-1 and 3.1-2.

The total collision frequency, ν, for the laboratory studies may be calculated as (Ref. G-37):

$$\nu = (3.5 \times 10^{22}/(MT)^{1/2})P_{mm} = 2.5 \times 10^{20} \text{ collisions/cm}^2 \text{ sec}$$

where P = 1 torr, T = 300 K, and M = 32 since the oxygen is predominantly as molecules. The number of collisions of oxygen atoms (10^{14}/cm^3) may be estimated from the number of total molecules at 1 atm at 300 K which is about 2.4×10^{19}/cm^3 then:

$$1 \times 10^{14}/2.4 \times 10^{19} = 4.1 \times 10^{-6} \text{ atm} \times 760 \text{ mm/atm} = 3.1 \times 10^{-3} \text{ mm}$$

and the collision frequency for the oxygen atoms becomes:

$$(2.5 \times 10^{20})(3.1 \times 10^{-3}) = 7.8 \times 10^{17} \text{ collisions/cm}^2 \text{ sec.}$$

From Table 3.1-8 for polyimide (Kapton) the weight loss is 4.1×10^{-7} g/cm^2 sec, so $4.1 \times 10^{-7}/7.8 \times 10^{17}$ = 0.5×10^{-24} g/oxygen atom. For Mylar it is $6.28 \times 10^{-7}/7.3 \times 10^{17}$ = 0.8×10^{-24} g/oxygen atom.

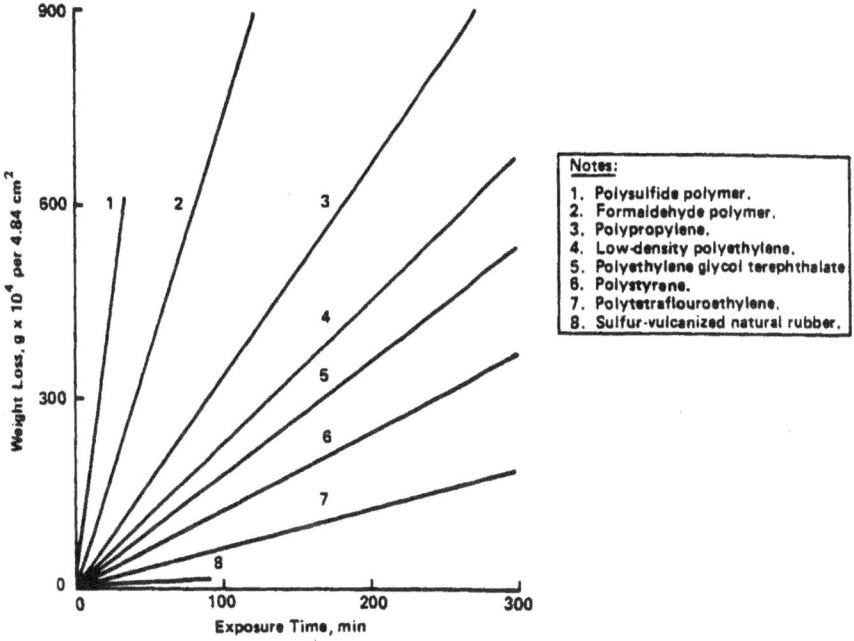

Figure 3.1-8 Relative Effects of Atomic Oxygen on a Variety of Polymers

Tables 3.1-1 and 3.1-2 show the flight R.E. for Kapton to range from (2-3) x 10^{24} cm^3/oxygen atom (taking the density of Kapton to be 1.4 g/cm^3), this is equivalent to (2.8-4.2) x 10^{-24} g/oxygen atom. For Mylar the R.E. ranges from (1.5-3.6) x 10^{-24} g/oxygen atom. The flight results for Kapton are about a factor of 4-6 greater than observed in the laboratory tests and about 2.5-6 times greater for Mylar. If the actual oxygen atom concentration was somewhat less than the estimated $10^{14}/cm^3$ in the laboratory tests, the agreement would be excellent. It appears that laboratory studies can give representative results (if the fluence in flight and concentration of oxygen atoms in the laboratory are known accurately), although the concerns discussed in Section 3.1.4 still apply.

It should be noted that the reaction efficiency gives essentially one point in time from which kinetics cannot be predicted. It is believed, however, that materials will undergo materials loss with "zero order" kinetics, i.e. linear in time. This is based upon the fact that no temperature effect was found in space exposed samples (Ref. G-4) and the results depicted in Figure 3.1-8 (Ref. G-33) where it can be noted that indeed weight loss is linear in time, albeit the time is much too short to be definitive. These results do, however, give veracity to the R.E. results listed in Table 3.1-1. Zero order kinetics are given by the equation:

$$dx/dt = k \qquad\qquad 3.1-6$$

where x is the weight loss (or surface recession of R.E.), t is time, and k is a constant. The constant k may be taken as the R.E. for a particular polymer. With the altitude of the spacecraft and a model atmosphere (see Section 4), the atomic oxygen fluence can be determined and a simple integration with time will give the weight (or thickness) loss for any given time.

The second, third and fourth entries of Table 3.1-10 are of particular importance for the space environment. It can be seen that polyethylene oxidation is enhanced by electron irradiation (energy not specified). The results have been plotted in Figure 3.1-9 from which the rate of oxidation as a function of dose can be obtained as:

$$\log (MRads) = \text{slope (Rate)} + \log \text{constant}$$

or

$$\text{Rate} = \frac{1}{\text{slope}} \log \left(\frac{MRads}{\text{const}}\right) \qquad 3.1\text{-}14$$

which for polyethlene is

$$\text{Rate} = 2.3 \times 10^{-7} \log \left(\frac{MRads}{5 \times 10^{-5}}\right) \qquad 3.1\text{-}15$$

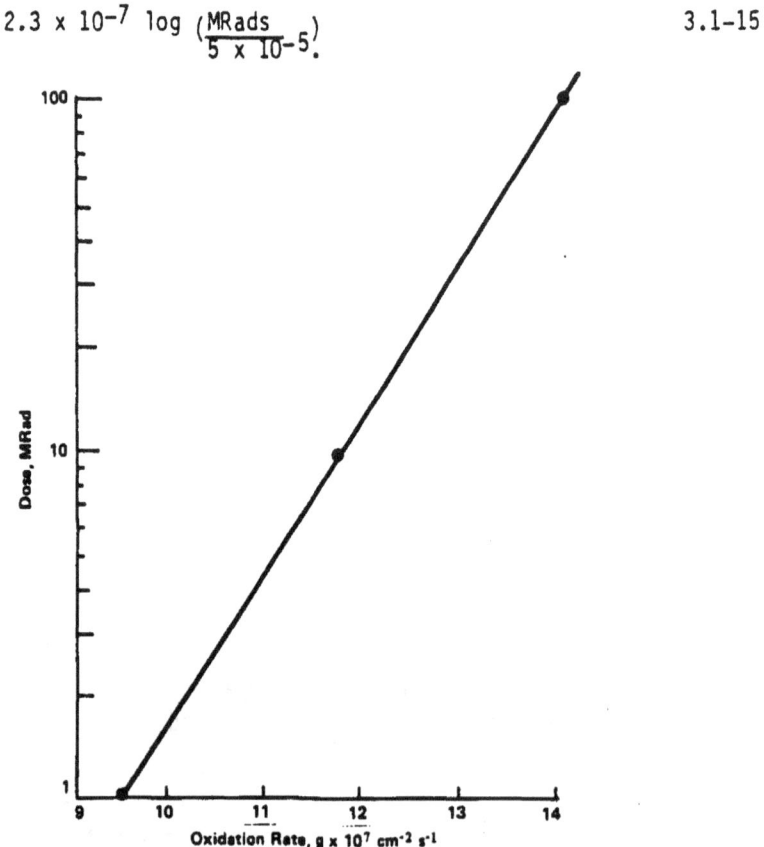

Figure 3.1-9
Oxidation Rate of Polyethylene as a Function of Electron Irradiation

Further laboratory testing of other polymer oxidation rates as a function of dose is very much warranted.

With zero order kinetics it is difficult to devise an accelerated laboratory test since the fluences of atomic oxygen required are beyond present technology. Therefore, it is suggested that on Shuttle flights, samples of a material be exposed for different lengths of time and the results be analyzed as a function of time. If, as expected, this is a linear function with time then the above approach will develop a data base for all those materials for which the reaction efficiencies are known.

This section has shown some of the complexities involved in interpreting the existing oxidation data base and indicates the caution which must be exercised when attempting to use the R.E. as a universal phenomena description.

3.1.5.2 Oxidation Effects Models

A Martin Marietta orbital oxidation model has recently been updated. The model has the capability of calculating oxygen fluence on spacecraft surfaces given arbitrary geometry. The Mass Spectrometer Incoherent Scatter (MSIS) atmosphere (see Section 4) is used to calculate oxygen density, which is integrated over the spacecraft orbit. The program then calculates surface recession rates for the materials identified for the various surface nodes. NASA JSC has a similar model as documented in Reference G-19.

3.1.6 Oxidation/Erosion Protection

Soon after the first observations of the deleterious effects of the LEO environment on materials, studies directed toward protection of materials were implemented.

In a review of oxidation/erosion on STS-1, -2, -3, -4, -5 and -8, it was noted that protection of organic films could be obtained by metal films of Al, Au, Pd, Pt, and Indium/Tin oxide (ITO) as well as with silicone coatings (Ref. G-19). Protective mechanism tests flown on STS-5 are listed in Table 3.1-11 (Ref. G-8). It can be noted that Kapton and Mylar, which show high reaction efficiencies (Table 3.1-2), can be adequately protected with an appropriate overcoat. A detailed study of Mylar by a variety of analytical tests such as Scanning Electron Microscopy (SEM), Fourier Transform Infrared (FTIR), and weight loss showed that while up to 75 percent of uncoated Mylar was eroded during exposure, thin coatings, on the order of 50Å, of Au/Pd and Al protected the material for oxygen fluences of at least 10^{21} O atoms/cm^2 (Ref. G-38).

Table 3.1-11 Materials Protection Test on STS-5 (Ref G-8)

Description	Properties of Interest	Results
Multilayered Anti-Reflection (Mylar) Coated Disc; Outermost Coating of ZnS	1. IR Transmission 2. Surface Effects	No Change S Removed and Replaced by O
Mylar-Coated Disc; Outermost Coating of ThF_4	1. IR Transmission 2. Surface Effects	No Change F Removed and Replaced by O
Mylar-Coated Disc; Outermost Coating of SiO	1. IR Transmission 2. Surface Effects	Slight Transmission Decrease No Detectable Change
In_2O_3/SnO_2-Coated SiO_2 (200Å, 400Å, 600Å)	1. Hall Conductivity 2. Thickness	No Change No Change
In_2O_3-Coated Optical Solar Reflector	1. UV-VIS-IR Reflectance 2. Hall Conductivity	No Change No Change
Siloxane (Owens-Illinois 100) Coated KRS-5	1. Surface Effects 2. IR Transmission	No Visible Effects, Enhanced Oxygen Concentration, Mass Loss Observed No Significant Change
Siloxane (GE RTV 560) Coated KRS-5	1. Surface Effects 2. IR Transmission	Surface Visibly Roughened, Enhanced Oxygen Concentration, Mass Loss Observed No Significant Change
ITO/Kapton	Weight Loss	No Change

Such tests were continued on STS-8. It was shown that the emissivity and absorptance of Kapton showed no changes when protected by DC6-1104 silicone (Ref. G-5). Sputtered nickel, chromium, and aluminum protected epoxy composites (Ref. G-15). A novel coating of metal oxides plus PTFE fluoropolymer was also tested. The reaction efficiencies are compared in Table 3.1-12 (Ref. G-39). Not only is there a significant protection against weight loss, the optical properties of Kapton remain essentially unchanged as can be noted by Figure 3.1-10.

Table 3.1-12
Mass Loss of Protected and Unprotected Kapton Samples to Low Earth Orbital Environment

Protective Coating on Kapton	Thickness of Protective Coating, Å	Mass Loss, Mg	Mass Loss per Incident Oxygen Atom*, g/Atom
None (Unprotected)	0	5020 ± 9.9	4.3×10^{-24}
Al_2O_3	700	587 ± 5.2	4.8×10^{-25}
SiO_2	650	5.9 ± 5.2	5.0×10^{-27}
96% SiO_2, 4% PTFE	650	10.3 ± 5.2	8.8×10^{-27}

*Based on an estimated atomic oxygen fluence of 3.5×10^{20} atoms/cm².

Protection tests were also conducted on STS-41G. Plasma sprayed FEP was found to provide protection for carbon-epoxy in some cases, but the effectivity was highly dependent on the integrity of the applied coating (Ref. G-25). Other coatings were examined in protection of paints and silver (Ref. G-40). OI-621 (otherwise unidentified), RTV-602 and MN41-1104-0 (silicone) were applied to Chemglaze Z302 specular paint, and MN41-1104-0 was applied to Chemglaze Z835. The MN41-1104-0 was found to be effective in protecting the Z835, although some darkening of the surface upon application occurred. Only the OI-651 was effective both in protecting the Z302 and preserving the specular properties of that surface. In protecting silver, aluminum, gold and palladium were applied to silver interconnects or foil. Protection was again highly dependent on the integrity of the coating. Gold was the only coating that did not show penetrations to the silver substrate, and this occurred only in 2 of 4 samples. All the gold samples showed discolorations due to oxygen exposure and some silver diffusion into the gold, indicating gold may be of limited effectiveness as a long duration protection technique (Ref. G-40).

Although most effects of orbital reactions with atomic oxygen appear detrimental, it has been suggested that orientation of contaminated surfaces into the velocity vector may be used to clean them (Ref. G-41).

If the protections discussed above cannot be carried out for any particular Shuttle experiment, an obvious approach could be to fly higher or not expose the experiment hardware to the ram direction.

3.1.7 Key Technical Personnel

	Organization	Phone
L. J. Leger	JSC	(713) 483-2059
J. T. Visentine	JSC	(713) 483-4564
A. F. Whitaker	MSFC	(205) 453-5975
M. Greenfield	NASA HQ	(202) 453-2862
L. E. Bareiss	Martin Marietta	(303) 977-8713
G. W. Sjolander	Martin Marietta	(303) 977-8686
G. S. Arnold	Aerospace Corp.	(213) 647-1935
D. R. Peplinski	Aerospace Corp.	(213) 648-6928
H. Garrett	JPL	(818) 354-2644
R. Liang	JPL	(818) 354-6314

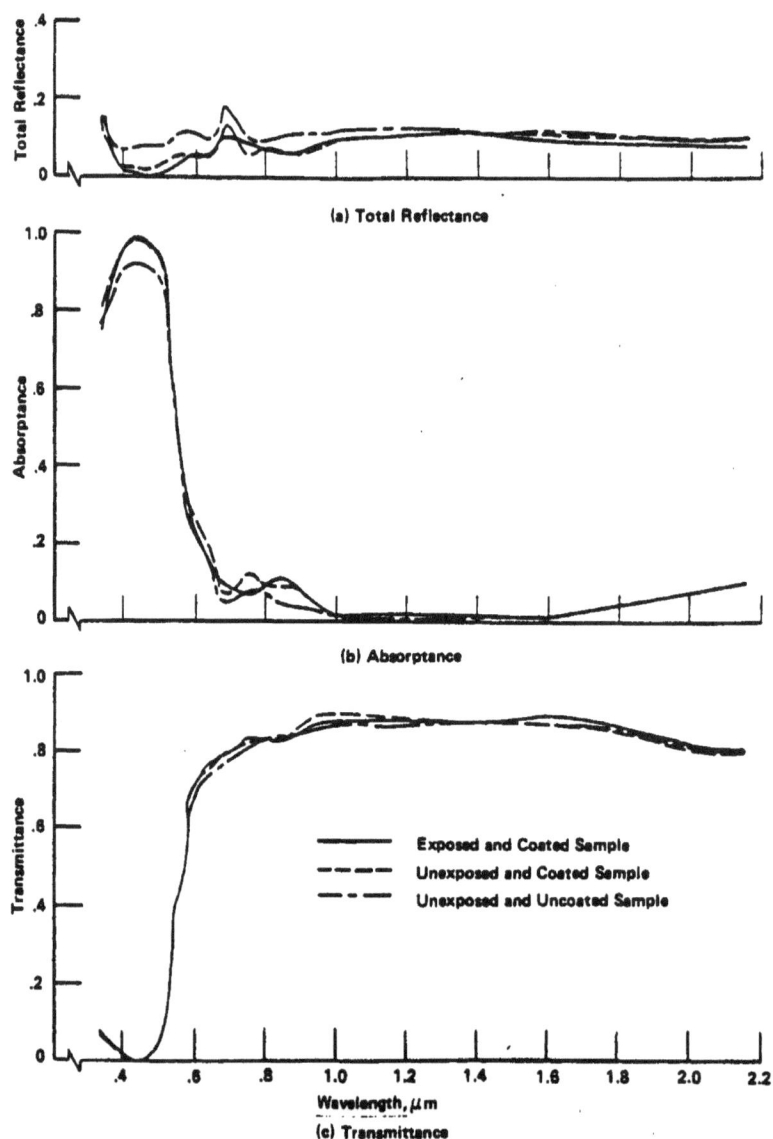

Figure 3.1-10
Optical Properties of 96% SiO_2 4% PTFE Coated Kapton Samples Unexposed and Exposed to Low-Earth Orbital Environment Compared with Uncoated and Unexposed Kapton

3.1.8 Oxidation References

G-1. Leger, L. J., "Oxygen Atom Reaction with Shuttle Materials at Orbital Altitudes," NASA TM-58246, May 1982.

G-2. Leger, L. J., "Oxygen Atom Reaction with Shuttle Materials at Orbital Altitudes Data and Experimental Status," AIAA 21st Aerospace Sciences Meeting, Reno, AIAA-83-0073, January 10, 1983.

G-3. Leger, L. J., "STS-8 Atomic Oxygen Experiment, Results Summary," Atomic Oxygen Working Group Meeting, Washington, D.C, January 23, 1984.

G-4. Visentine, J. T., L. J. Leger, J. F. Kuminecz, and I. K. Spiker, "STS-8 Atomic Oxygen Effects Experiment," AIAA 23rd Aerospace Sciences Meeting, Reno, AIAA-85-0415, January 14, 1985

G-4A. Gregory, J. C. and P. N. Peters, "The Production of Glow Precursors by Oxidative Erosion of Spacecraft Surfaces", in NASA Conference Publication 2391, Second Workshop on Spacecraft Glow, May 6-7, 1985.

G-5. Park, J. J., "EOIM GSFC Materials," Atomic Oxygen Working Group Meeting, Washington, D.C., January 23, 1984.

G-6. VanLoon, S., Martin Marietta Memorandum 84-0787M/0019M, January 11, 1984.

G-7. Whitaker, A. F., S. A. Little, R. J. Harwell, O. B. Griner, and R. F. DeHaye, "Orbital Atomic Oxygen Effects on Thermal Control and Optical Materials STS-8 Results," AIAA 23rd Aerospace Sciences Meeting, Reno, AIAA-85-0416, January 14, 1985.

G-8. Stuckey, W. K., E. N. Borson, M. Meshishnek, E. Zehms, G. S. Arnold, P. A. Bertrand, D. F. Hall, M. S. Leung, D. R. Peplinski, and L. U. Tolentino, "Effects on Optical and Metallic Surfaces," AIAA Shuttle Environment and Operations Meeting, Washington, D.C., October 31, 1983.

G-9. Park, J. J., T. R. Gull, H. Herzig, and A. R. Taft, "Effects of Atomic Oxygen on Paint and Optical Coatings," AIAA-83-2634-CP, October 1983.

G-10. Stuckey, W. K., "STS-8 EOIM Experiment," Atomic Oxygen Working Group Meeting, Washington, D.C., January 23, 1984.

G-11. Whitaker, A. F., "MSFC STS-8 Materials Evaluations," Atomic Oxygen Working Group Meeting, Washington, D.C., January 23, 1984.

G-12. Slemp, W. S., B. Santos-Mason, G. F. Sykes, Jr., and W. G. Witte, Jr., "Effects of STS-8 Atomic Oxygen Exposure on Composites, Polymeric Films and Coatings," AIAA 23rd Aerospace Sciences Meeting, Reno, AIAA-85-0421, January 14, 1985.

G-13. Leger, L. J., I. K. Spiker, J. F. Kuminecz, T. J. Ballentine, and J. T. Visentine, "STS Flight 5 LEO Effects Experiment Background Description and Thin Film Results," AIAA Shuttle Environment and Operations Meeting, Washington, D.C., AIAA-83-2631-CP, October 31, 1983.

G-14. Peters, P. N., R. C. Linton, and E. R. Miller, "Results of Apparent Oxygen Reactions on Ag, C, and Os Exposed during the Shuttle STS-4 Orbits," Geophys. Res. Lett., 10, 569, 1983.

G-15. Slemp, W. S., "Preliminary Results of LARC Studies on Materials Exposed in the STS-8 Atomic Oxygen Flight Experiment," Atomic Oxygen Working Group Meeting, Washington, D.C., January 23, 1984.

G-16. Misra, M. S., "Interaction of Atomic Oxygen and Silver Interconnects for STS-8 Flight Experiments," Martin Marietta MCR-43-551, April 1984.

G-17. Gull, T. R., H. Herzig, and A. R. Taft, "Evaluation of Oxygen Interaction with Materials," Atomic Oxygen Working Group Meeting, Washington, D.C., January 23, 1984.

G-18. Triolo, J., R. McIntosh, R. Kruger, and N. Pugel, "Goddard Space Flight Center Atomic Oxygen Monitor," Atomic Oxygen Working Group Meeting, Washington, D.C., January 23, 1984.

G-19. Leger, L. J., J. T. Visentine, and J. A. Schliesing, "A Consideration of Atomic Oxygen Interactions with Space Station," AIAA 23rd Aerospace Sciences Meeting, Reno, AIAA-85-0476, January 14, 1985.

G-20. Park, J. J., "Results of Examination of Material from the Solar Maximum Recovery Mission", proceedings of the SMRM Degradation Study Workshop, May 1985.

G-21. Liang, R. H., K. I. Oda, and S. Y. Chung, "Degradation Studies of SMRM Teflon", Proceedings of the SMRM Degradation Study Workshop, May 1985.

G-22. Linton, R., and A. Whitaker, "SSM Atomic Oxygen Reactions on Kapton and Silverized Teflon", Proceedings of the SMRM Degradation Workshop, May 1985.

G-23. Stuckey, W. K., A. A. Gluska and J. Uht, "Analysis of Normal and Transparent Silver Teflon", Proceedings of SMRM Degradation Study Workshop, May 1985.

G-24. Santos-Mason, B., "Preliminary Results of the SMM Exposed Aluminized Kapton and Silvered Teflon", Proceedings of the SMRM Degradation Study Workshop, May 1985.

G-25. Zimcik, D. G. and C. R. Maag, "Results of Apparent Atomic Oxygen Reactions with Spacecraft Materials During Shuttle Flight STS-41G", AIAA-85-7020, AIAA Shuttle Environment and Operations II Conference, November 1985.

G-26. Leger, L. J., private communication.

G-27. Leger, L. J., "EOIM-3 Working Group Meeting Minutes, December 10-11, 1985", NASA JSC.

G-28. Arnold, G. S., and D. R. Peplinski, "Reaction of High Velocity Atomic Oxygen with Carbon," AIAA 22nd Aerospace Sciences Meeting, AIAA-84-0549, January 10, 1984.

G-29. Arnold G. S., and D. R. Peplinski, AIAA Journal, to be published.

G-30. Park, C., "Effects of Atomic Oxygen on Graphite Ablation," AIAA Journal, 14, 1649, 1976.

G-31. Redhead, P. A., J. P. Hobson, and E. V. Kornelsen, The Physical Basis of Ultra High Vacuum, Chapman and Hall, Ltd., London, 1968.

G-32. Arnold, G. S., and D. R. Peplinski, "Reaction of Atomic Oxygen with Polyimide Films," AIAA Journal, accepted for publication.

G-33. Hansen, R. H., J. V. Pascale, T. De Benedicts, and P. M. Rentzepis, "Effect of Atomic Oxygen on Polymers", J. Polym. Sci. (A), 3, 2205, 1965.

G-34. Miller, W. L., "Mass Loss of SHuttle Space Suit Orthofabric Under Simulated Ionospheric Atomic Oxygen Bombardment", NASA, TM-87149, November 1985.

G-35. 13th Space Simulation Conference, NASA Conference Publication 2340, 1984.

G-36. Bareiss, L. E. and G. W. Sjolander.

G-37. Glasstone, S., Textbook of Physical Chemistry, Van Nostrand Co. Inc., New York, 1946.

G-38. Fraundorf, P., D. Linstrom, N. Pailer, S. Sanford, P. Swan, R. Walker and E. Zinner, "Erosion of Mylar and Protection by Thin Metal Films," AIAA Shuttle Environment and Operations Meeting, Washington, D.C., AIAA-83-2636, October 31, 1983.

G-39. Banks, B. A., M. J. Mirtick, S. K. Rutledge, and D. M. Swec, "Sputtered Coatings for Protection of Spacecraft Polymers," NASA TM-83706.

G-40. Whitaker, A. F., J. A. Burka, J. E. Coston, I. Dalins, S. A. Little and R. F. DeHaye, "Protective Coatings for Atomic Oxygen Susceptible Spacecraft Materials--STS-41G Results", AIAA-85-7017, AIAA Shuttle Environment and Operations II Conference, November 1985.

G-41. Peters, P. N., and E. R. Miller, NASA-MSFC Suggestion 79-49, 1979.

3.2 SURFACE GLOW PHENOMENA

3.2.1 Introduction

Nighttime photographs taken by the crew of STS-3 revealed a reddish glow originating on and above the vehicle surfaces exposed to the ram direction (Ref. H-1). Similar emissions were found on earlier satellites, including unexplained enhancements in the airglow measured by the Vehicle Airglow Experiment (VAE) on the Atmospheric Explorer-C (AE-C) Satellite (Ref. H-2).

All subsequent STS flights have shown glows similar to the original STS-3 observations. Further, it has been observed that there is a prompt enhancement of the glow associated with thruster engine firings. A recent STS flight carried the Imaging Spectrometric Observatory (ISO). At no time did the instrument make direct observations of Orbiter surfaces. Looking directly into the velocity vector, the instrument detected optical effects associated with the high altitude ambient atmosphere as well as emissions occurring in the vicinity of the Orbiter due to the local environment (Ref. H-3). As with the red glow associated with the vehicle surfaces, a major feature of the ISO data appears to be molecular bands which lie in the 6000Å to 8000Å wavelength region of the red glow.

An experiment flown on STS-9/SL-1 provided evidence that the Shuttle glow extends into the ultraviolet from 1300Å to 1800Å (Ref. H-4), while other SL-1 and SL-2 instruments further recorded visible light and infrared glow phenomena.

3.2.2 Available Data

On the AE satellites, spectral measurements showed that above 160 km altitude the glow brightness at 6563Å and at 7320Å decreased with altitude at the same rate as the atomic oxygen number density. Therefore, it was suggested that the emissions were produced by the same mechanism, with atomic oxygen having an important role (Ref. H-5 and H-6). Below 160 km the glow brightness increased greatly with decreasing altitude and was no longer proportional to the atomic oxygen number density. Figure 3.2-1 depicts both the emission and oxygen number densities as a function of altitude. Figure 3.2-2 shows the spectral variation of the brightness. Although there was a different dependence of glow brightness in the two altitude regions, from the similarity of the brightness variation it was suggested that the glow, in both altitude regions, is produced by the same mechanism, with atomic oxygen being important in both regions.

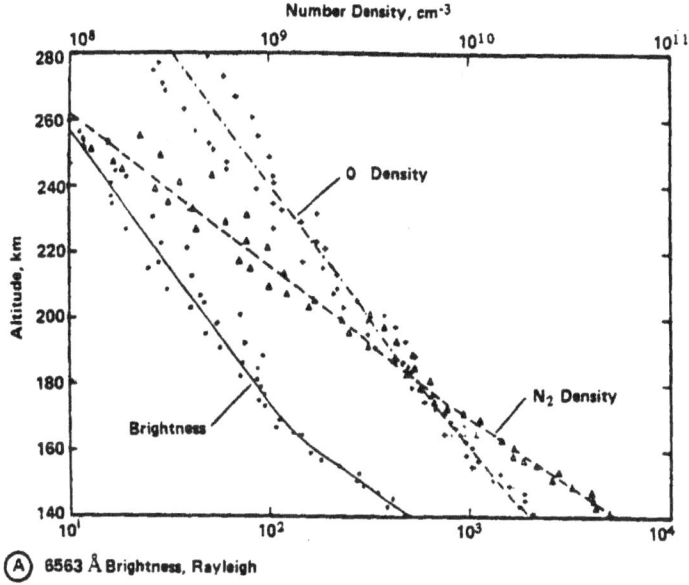

(A) 6563 Å Brightness, Rayleigh

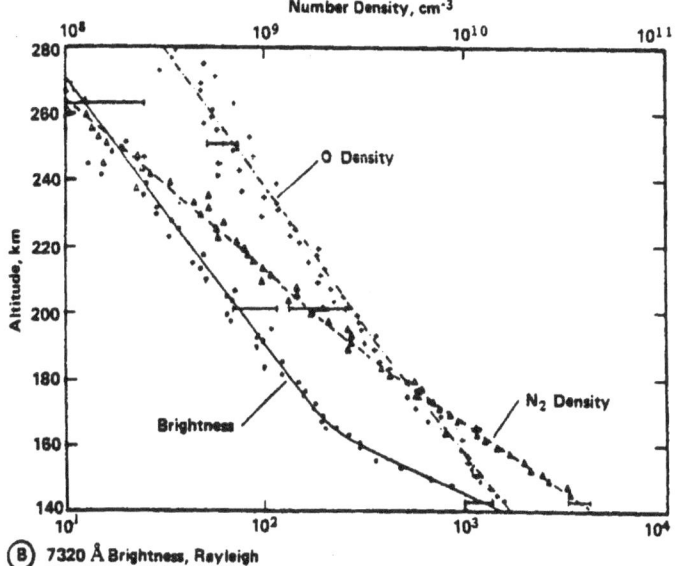

(B) 7320 Å Brightness, Rayleigh

Figure 3.2-1
The Altitude Variation of the Glow Emission at (a)
6563 Å and (b) 7320 Å, Along Measured Number
Densities of Atomic Oxygen and Molecular Nitrogen
(Ref H-5)

A Fabry-Perot (F-P) interferometer was flown on the Dynamics Explorer-B satellite (DE-B). It measured the density of metastable $O(^1S)$ and $O(^1D)$ atoms and the $O^+(^2P)$ ion. The measurements were made with a high resolution F-P etalon which performs a wavelength analysis on light detected from atmospheric emission features by spatially scanning the interference fringe plane with 12 concentric ring detectors. The scan is linear in wavelength covering a spectral range equal to 0.01796Å per detector channel at 7320Å. The spectral region for analysis is selected by a 10 Å half width interference filter centered at 7320Å. It was suggested that the glow spectrum compared favorably with the nightglow $OH(X\ ^2\Pi)$ spectrum suggesting it to be one of the species producing the glow (Ref. H-7). It had been pointed out that the OH airglow in the spectral region around 6563Å and 7320Å had the same relative intensity as the glow on the AE satellites (Ref. H-8). Further support for the OH emission is provided by a more extensive spectral range comparison (Ref. H-9).

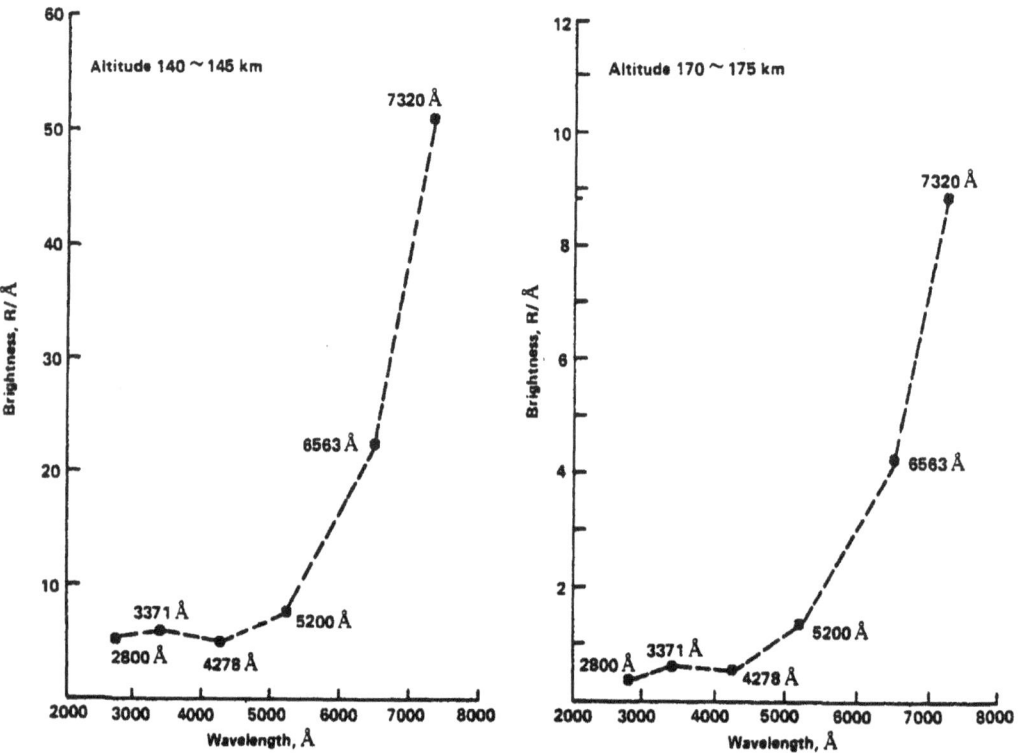

Figure 3.2-2
The Spectral Variation of the Glow Emission on AE-C at (a) 140-145 km and (b) 170-175 km (Ref H-5).

The Shuttle STS-3 nighttime photograph of the glow from the vertical stabilizer and engine pods was analyzed showing the intensity of the glow to be maximized about 20 cm off of the surfaces (Ref. H-10). The lifetime of the excited species was determined to be 0.67 msec. Comparison of the surface glow intensity with the 5577Å airglow in the photograph background yielded a surface glow intensity of 30 kR (R = Rayleigh = 10^6 photons/cm^2 sec).

On STS-4 a spectral determination was made showing the glow to extend from 6000Å to 8000Å. Comparison of photographs from STS-3 (240 km) and STS-5 (305 km) showed the intensity of glow to be brighter on STS-3 by a factor of 3.5 (Ref. H-11). On STS-4 the glow was measured with a transmission grating mounted in front of a photographic camera, and several exposures were taken on-orbit to make preliminary spectral measurements. The glow observed was predominantly in the 6300-8000Å range of the instrument band pass of 4300-8000Å (Ref. H-11). On STS-5 a Noctron 5 Image Intensifier was used. This was inserted between the body of a Nikon 35mm camera and the lens (55mm, f/1.4); for the starboard aft flight deck window a conventional Hasselblad camera (100mm, f/3.5) was used (Ref. H-12). As with the AE satellites (Ref. H-7), as the angle of incidence between the spacecraft surface and the velocity vector decreases the glow increases.

Spectral analysis of the results obtained on Spacelab 1 showed that the earlier suggestion (Ref. H-7) of OH as the candidate specie for producing the glow on Shuttle is probably not correct (Ref. H-13). The 5577Å of O(I) and the O_2 atmospheric (0,0) band at 7620Å were clearly identified while the OH Meinel bands were absent. With the elimination of these bands by the instrumentation, Figure 3.2-3 shows the remaining structureless red glow of the Shuttle (H-14).

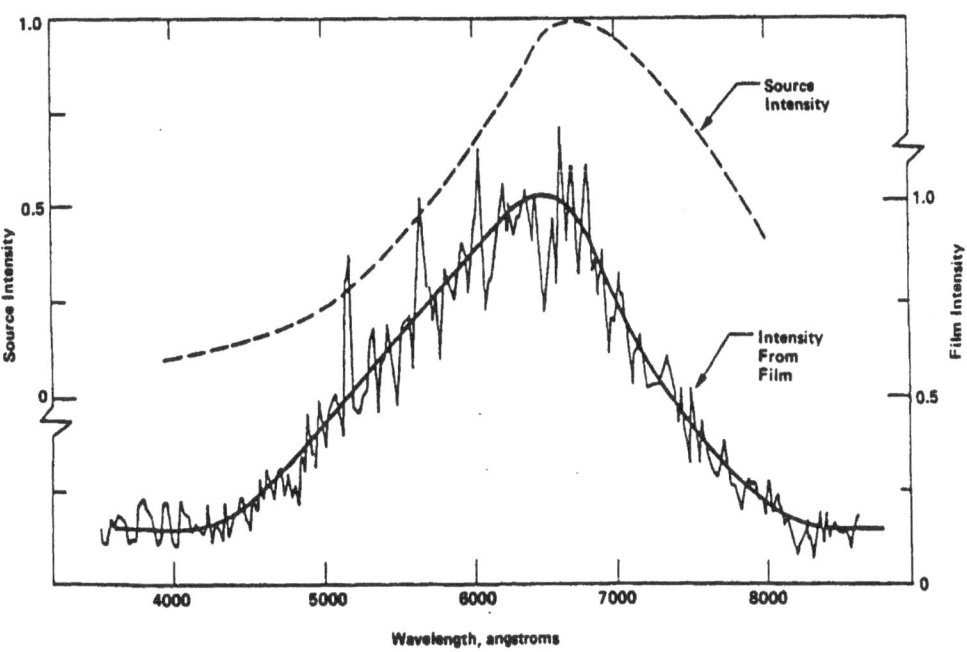

Figure 3.2-3
(Bottom) Six line average tracing of spacecraft glow. The tracing has been corrected for the calibrated D-Log-E response of the film. The noise character in the data is primarily because of ion scintillations in the image intensifier that have accumulated in the image over the 30-second exposure. (Top) Corrected spectrum of spacecraft glow where the instrument response has been applied to the curve drawn through the data shown in the bottom of the figure (Ref H-14).

Project FAUST, also on Spacelab 1, provided the first evidence that the Shuttle glow may extend into the 1300Å to 1800Å far UV region (Ref. H-4). The FAUST telescope provides wide field (8° diameter) imaging in the far ultraviolet (1300-1800Å) and was located on a pallet in the cargo bay. It employs a Wynne optical configuration to image the wide field onto a flat focal plane. The image falls on a frequency converting image intensifier tube which transforms the ultraviolet image into an intensified optical image which is then recorded on a 130a-0 spectroscopic film. The CaF_2 window of the telescope provides the short wavelength cutoff, and the falling sensitivity of the CsI on the frequency converter tube provides the long wavelength cutoff. These combine to give a 500Å bandpass with a maximum sensitivity at 1450Å. Preflight calibration showed the instrument could detect a 17.5 magnitude source in ten minutes against a dark field assuming an A0 stellar source. The glow in the UV was established by the correlation at the 80 percent confidence level between the background intensity and the angle between the view direction and the velocity vector (Ref. H-4).

The intensity of the glow from a variety of materials flown on the STS-41D RMS was assessed (Ref. H-15). The samples were ranked from 1 to 9 in order of their glow. Polyethylene was assigned 1, having the dimmest glow. The low signal to noise ratio made it difficult to draw any strong conclusions, but, the results are presented in Table 3.2-1.

Table 3.2-1
Material Glow Intensities (Arbitrary Units)

Material	Ranking
Z302 (Overcoated with Si)	9
MgF_2	8
Z302	7
Z306	6
Chemical Conversion Film	5
Carbon Cloth	4
Anodized Aluminum	3
401-C10	2
Polyethylene	1

The explanation of prompt enhancement of the glow surrounding the Orbiter following thruster firings remains open (Ref. H-16). However, studies made primarily to study the F region of the ionosphere appear to have relevance. An ionospheric "hole" can be created by chemical reactions of reactive molecules with the ionospheric plasma (Ref. H-17). Highly reactive molecules (such as H_2O, H_2 and CO_2) exhausted by a rocket engine into the ionosphere at altitudes, $h \geq 200$ km, where O^+ is the dominant ion, causes transformation to molecular ions at rates 100 to 1000 times faster than those occurring with the environmental N_2 and

O_2. These reactions are:

$$O^+ + H_2O \xrightarrow{k_1} H_2O^+ + O \qquad k_1 = 2.4 \times 10^{-9} \text{ cm}^3/\text{sec}$$

$$O^+ + H_2 \xrightarrow{k_2} OH^+ + H \qquad k_2 = 2.0 \times 10^{-9} \text{ cm}^3/\text{sec}$$

$$O^+ + CO_2 \xrightarrow{k_3} O_2^+ + CO \qquad k_3 = 1.2 \times 10^{-9} \text{ cm}^3/\text{sec}$$

Once formed, these molecular ions undergo rapid dissociative recombination with ambient electrons.

$$H_2O^+ + e^- \xrightarrow{k_4} OH^* + H \qquad k_4 = 3 \times 10^{-7} \text{ cm}^3/\text{sec}$$

$$OH^+ + e^- \xrightarrow{k_5} H + O^* \qquad k_5 = 1 \times 10^{-7} \text{ cm}^3/\text{sec}$$

$$O_2^+ + e^- \xrightarrow{k_6} O + O^* \qquad k_6 = 2 \times 10^{-7} \text{ cm}^3/\text{sec}$$

This leads to the loss of ion-electron pairs or the hole. "The creation of an ionospheric hole via the reactions described above will be accompanied by a significant amount of airglow emissions." (Ref. H-17). These include emissions from excited states of atomic oxygen at 5577Å and 6300Å. In support of these comments, it should be noted that the density monitor on Shuttle showed plasma depletions during thruster firings as large as a factor of ten (Ref. H-18).

Emissions from the vehicle environment (for example, from measurements made looking away from the Shuttle and Earth into the velocity vector and the high altitude dayglow) have been observed. The instrument used comprised an array of five spectrographs, each covering a portion of the wavelength range from 300Å to 12,700Å. Each spectrograph contained a focal plane detector in the form of an intensified two-dimensional charge coupled device array. The five spectrometers operated in parallel, each imaging approximately 200Å simultaneously. The full wavelength range covered by each spectrograph could be obtained in 20 steps. The array of spectrometers is called the Imaging Spectrometric Observatory (ISO). The observed emissions have the broad spectral characteristics of the N_2 First Positive system in the region of 6000Å to 8000Å long wavelength limit of the data obtained (Ref. H-19). If these are due to this system they are considerably brighter than would be expected by comparison with the N_2 Second Positive system. The enhancement of these red bands appears to be present in data taken with the field-of-view looking in directions other than the velocity vector, and thus are from the vehicle environment and not from surfaces within the instrument.

3.2.3 Additional Glow Data

3.2.3.1 Unpublished Glow Data

The IRT of SL-2 experienced high background signal levels throughout the SL-2 mission. Evaluation of these signals, including their relationship to glow, have not been published. In conjunction with the IRT experiment, ground based IR observations of SL-2 were also made. These results are also currently unpublished.

3.2.3.2 Needed Glow Data

The most important needed data on the Shuttle glow is a complete, high resolution spectrum of the glow, from IR to UV wavelengths, and taken from the PLB to avoid Shuttle window interference with data. Also needed are ground simulators to verify proposed mechanisms and acquire a materials response database.

Since the ionospheric hole discussed above is a viable approach to the understanding of the prompt enhancement of glow during thruster firings, it would be of great interest to determine its decay rate and spatial distribution.

3.2.3.3 Planned Glow Data

Among the planned Shuttle glow experiments is the Shuttle Infrared Glow Experiment (SIRGE), a low resolution liquid nitrogen cooled filter wheel photometer to be flown on a Hitchhiker-G getaway special (GAS). The spectrometer will cover 0.9 to 5.5 µm wavelengths, with a resolution ($\lambda/\Delta\lambda$) of 100 (Ref. H-20). Also planned is the flight of a UV spectrometer, covering 1900 to 3000 Å, This instrument is also a GAS can PLB experiment (Ref. H-21).

In addition to the above planned experiments, several other experiments have been proposed, including those of References H-22, H-23 and H-24.

3.2.4 Ground Simulation Studies

The technical literature is replete with studies of glows from ions (generally high energy) impinging on metals, but there is a paucity of laboratory studies related to the glow from Shuttle related materials.

One published study of low energy oxygen ions impinging on Kapton, film has shown a faint white glow extending about 5 mm in front of the impacted surface, and behind the sample holder a distinct greenish tinge could be seen in the diffuse glow of the beam (Ref. H-20). The white glow was attributed to continuum radiation from oxygen recombination at the surface, and the green glow appeared spectroscopically to be from the First Negative bands of O_2^+.

Some very preliminary studies of the impingement of ions on Chemglaze Z306 are worth consideration (Ref. H-26). A modified Colutron ion beam gun delivered ions ranging from 500 to 1000 eV. An EMR phototube, with a range between 4000 and 9000Å, was placed in the vicinity of the target and recorded the light output. The experiment was then repeated with a cutoff filter that passed light of wavelengths greater than 6000Å. Figure 3.2-4 shows the ratio "with filter/without filter" after the appropriate normalizations. It can be seen that all ions appear to give a red glow; however, it should be noted that these ion energies are far above

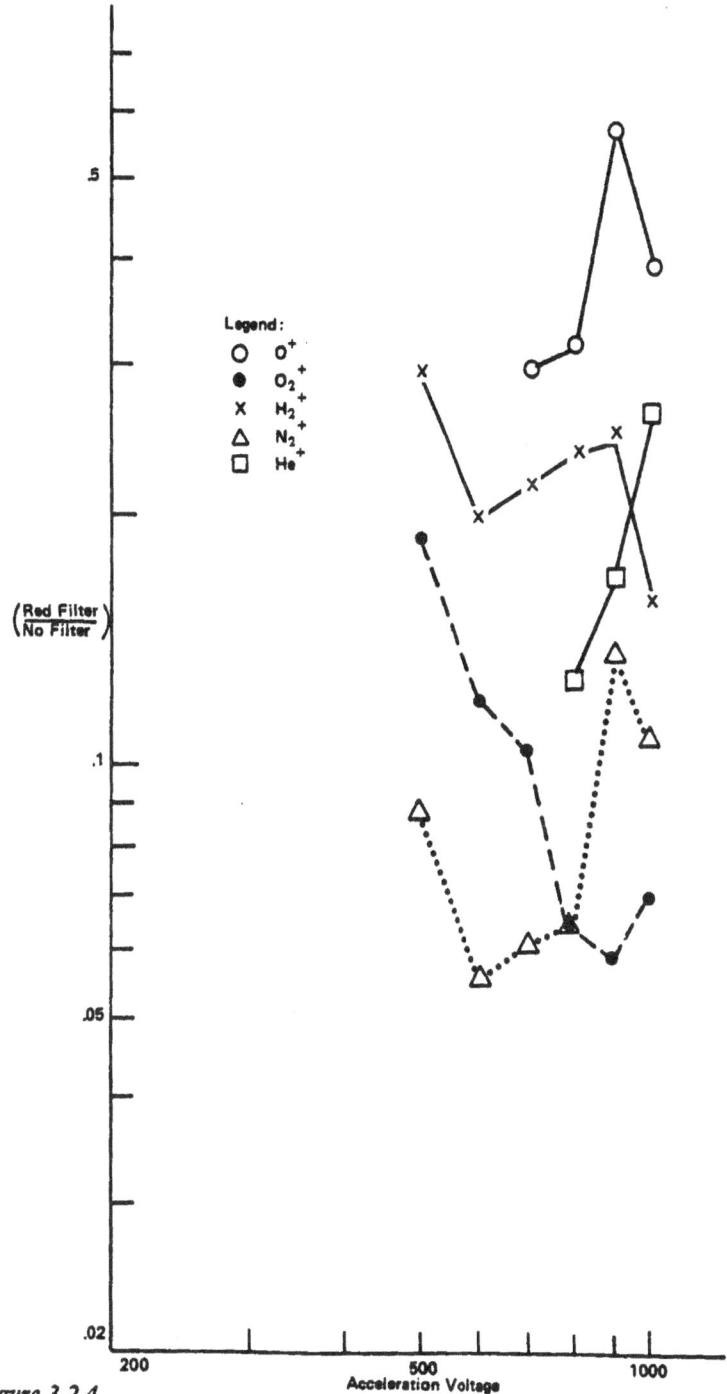

Figure 3.2-4
Percentage of Glow Longer than 6000Å vs Acceleration Voltage of Ions

those impinging on Shuttle. After further upgrades to the ion gun system, it is planned to study these ions and appropriate neutrals at much lower energies.

3.2.5 Models for the Shuttle Glow

A variety of processes have been suggested to account for the surface glow. Among these are:

a. Impact excitation of species in the direction of the velocity vector.

b. Excitation of absorbed species, both atomic and molecular.

c. Sputtered species either excited or subsequently excited by reaction with the vehicle atmosphere.

d. Luminescence of the solid surfaces.

e. Surface recombination of atomic and molecular species.

f. Surface recombination of ions and electrons.

These processes are followed by emission from excited entities. The suggested processes listed above are not discussed in detail.

The glow of the AE satellite was suggested to be produced by the chemiluminescent combination of O and NO (Ref. H-2). However, the measurements at 7320Å and 6563Å did not coincide with laboratory measurements of the spectral distribution from the NO-O reaction. Therefore this is considered to be an unlikely process (Ref. H-5). The spectral measurements on the DE-B satellite provides considerable evidence for OH as one of the species producing the glow on such spacecraft (Ref. H-6). Figure 3.2-5 shows a comparison of the measured OH nightglow and the spacecraft glow spectra (the shaded area indicates the statistical uncertainty in the measurements). The shape of the glow spectrum from channels 4 through 12 agrees with that of the OH spectrum. Thus OH might be one of the species producing glow on satellites.

From the similarity of spectra between Shuttle glow and the continuum chemiluminescent reaction of NO with atomic oxygen depicted in Figure 3.2-6 (developed from References H-27, H-28 and H-29), the NO model for the glow is being reconsidered. The mechanism is depicted in Figure 3.2-7 (Ref. H-30). No explanation has been given for the so-called atmospheric atomic nitrogen which begins the entire sequence by interaction with atomic oxygen on the Shuttle surface. Until a source for atomic nitrogen is elucidated, this mechanism must be considered speculative.

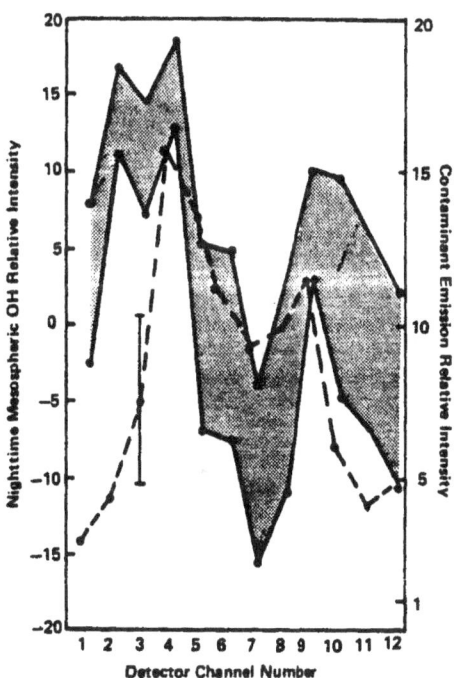

Figure 3.2-5
Comparison of the OH Nightglow (Brokenline) and the Comtaminant Glow Spectrum (Ref H-6)

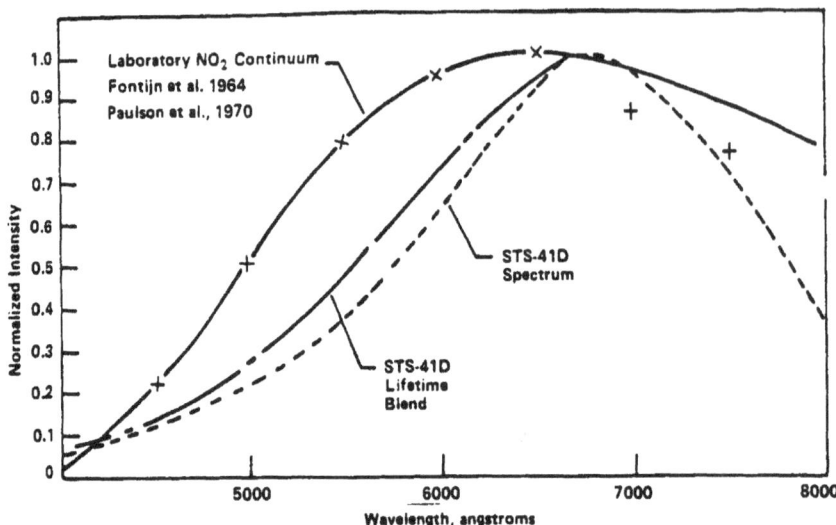

Figure 3.2-6
The spectrum of spacecraft glow compared with that of the laboratory spectrum measured in laboratory experiments by Fontijn et al. (Ref H-27) and Paulsen et al. (Ref H-28). A spectral blend produced by spectrally e-folding the measured spectrum with lifetime data of Schwartz and Johnston (Ref H-29) is also plotted (Ref H-14).

An analysis of the glow photograph of the vertical stabilizer on STS-3 determined the lifetime of the excited entity to be 0.67 msec. (Ref. H-10). This is almost an order of magnitude shorter than the OH $(X,^2\Pi)$ radiative lifetime and suggests that some other species besides OH is producing the Shuttle glow. An estimate of the glow intensity of 30 kR for a sight column intensity corresponds to a maximum volume emission rate of 7 $\times$ 10^6 photons/cm^3 sec. In turn this corresponds to a number density of 4.7 $\times$ 10^3/cm^3 at the surface. The incoming flux of atomic oxygen is about 1.4 $\times$ 10^{15}/cm^2 sec so that the photon production efficiency is 10^{-7} excited molecules per impacting oxygen atom. The efficiency could lie higher if emission is present at wavelengths other than that recorded by the photographic emulsion. As with the AE satellites, the Shuttle glow is brighter at the lower altitudes suggesting an association of the glow with atomic oxygen (Ref. H-12). The best estimate of the glow intensity ratio between STS-3 (240 km) and STS-5 (305 km) glow is about 3.5 and is consistent with the decrease in atomic oxygen density. The glow is observed predominantly in the region 6400Å to 8000Å of the instrumental band pass of 4300Å to 8000Å. It should be noted here that the AE satellite data (Figure 3.2-2) shows emission down to 2800Å. This along with the suggestion that the glow on Shuttle arises from bands similar to the atmospheric O_2 bands indicates that the glows on the smaller satellites (AE) are different from those of the much larger Shuttle. Furthermore, the AE spectral radiances (Ref. H-4) at 240 km indicate total column emission rates of only 1.5 kR at wavelengths less than 7330Å whereas those from STS-3 (240 km) were estimated as 10 to 100kR (Ref. H-31). The order of magnitude longer path lengths in the Shuttle Orbiter's viewing projection are not sufficient to account for its approximately two orders of magnitude higher radiance.

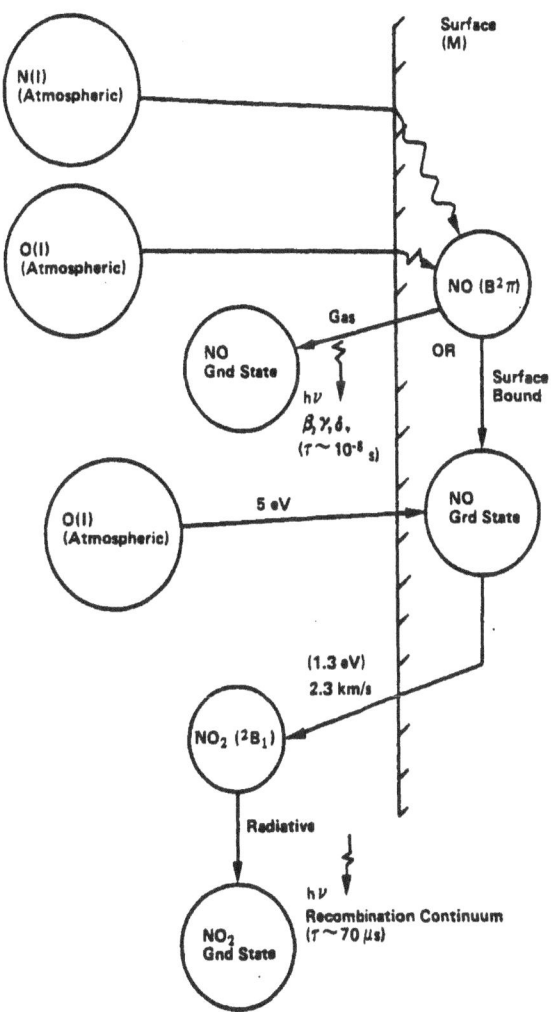

Figure 3.2-7
A schematic representation of the chemistry believed to be responsible for spacecraft ram glow. Starting at the top, the ramming O(I) and N(I) intercept a spacecraft surface and form NO, some of which sticks to the surface and some of which escapes in the gas phase. The NO that sticks to the surface is subjected to ramming O(I) which forms a 3-body recombination with the surface (M) to create NO_2. The escaping NO_2 radiates the red continuum observed on ram surfaces (H-22).

Plasma interaction calculations show that about 10^{10} eV/cm^3 sec to be deposited by low energy electrons ~10 eV per electron (Ref. H-32). If the glow from the Shuttle is taken as 30 kR this amounts to about 7 x 10^6 photons/cm^3 sec. Taking an average photon energy of 1.7 eV

3-43

(~7000Å) about 1.2×10^7 eV/cm^3 sec appears in emission for an efficiency of 1.2×10^{-3}. This is some four orders of magnitude greater than the calculated efficiency for STS-3. It has been suggested that the plasma interaction model also fails because the emission spectrum of the glow differs substantially from what would be expected from electron impact spectra on a mixture of O, N_2, O_2 and H_2O (Ref. H-33). However, energetic electron impact on a mixture of N_2/O_2 is shown in Figure 3.2-8 from which it can be seen that the molecular nitrogen First Positive system appears in the appropriate wavelength range and increases in intensity toward the red (Ref. H-34).

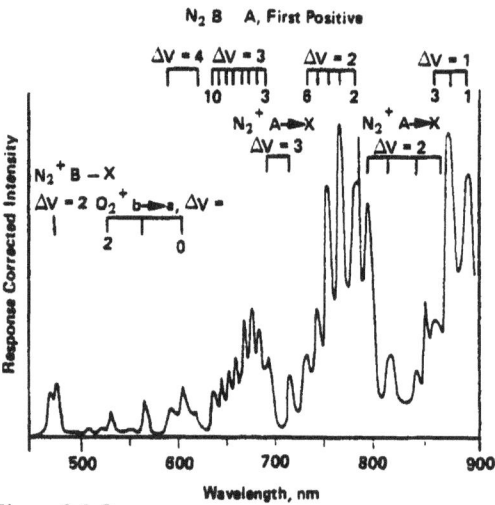

Figure 3.2-8
Emission from Electron-Irradiated Air at 0.3 mtorr Pressure, Dominated by N_2 First Positive Emission which Rises to the Red (Ref H-29).

Another process which can also lead to emission of the First Positive system of molecular nitrogen is the recombination of N atoms on Shuttle surfaces formed from the collision induced dissociation of N_2 (Ref. H-29). The surface will incompletely accommodate the 9.8 eV (dissociation energy) that becomes available and the molecules will leave the surface, some fraction of which will be in high vibrational levels of the electronically excited A $^3\Sigma_u$ state. These then decay to lower vibrational levels of the B $^3\Pi_g$ state (a reverse First Positive transition). Then the decay B⟶A leads to the emission of the First Positive system of molecular N_2.

However appealing, this process has some difficulties. The radiative lifetimes for the A⟶B transitions are quite slow, being on the order of 2.5 sec for the v = 8 level of the A state and about 3 msec for the v = 20 level (Ref. H-30). Since the model postulates surface recombination, the molecules leave the surface at thermalized velocities. If 300 m/sec is assumed for this velocity the molecules will be on average 7.5×10^4 cm away (for v = 8) or 90 cm away (for v = 20) from any surface before the B⟶A transitions occur. The glow appears to peak at 20 cm from the surface and is in the noise at 40 cm from the surface (Ref. H-10). Furthermore, the energy pooling into the B state requires the collision of A state molecules with another such A or excited ground state ($N_2^*(X,{}^1\Sigma_g^+)$) molecule (Ref. H-35 and H-36). This evaluation neglects, however, collisions with the ambient atmosphere, which could serve to concentrate molecules near the surface. In another possible approach, if the surface recombination leads directly to the B state, there is the same difficulty with lifetime (2-8 μsec) which leads to emission directly on the surface and not some 20 cm away. Resolution of these lifetime speculations are still uncertain. Testing or modeling is needed.

The emission from OH has been suggested for the glow emission on the AE satellites. Dissociative collisions of H_2O with the Shuttle surfaces could produce OH in high vibrational states and could result in the orange-red component. The dissociation energy of H from H_2O requires only 3.07 eV leaving on the order of 2 eV for vibrational excitation leading to Meinel emissions. However, these exhibit spectral structure which should have been partially resolved in the observation (Ref. H-37).

The chemiluminescent combination of O and NO was discounted as a possibility because of a lack of matching with the observed glow (See above). Another nitrogen oxide has been suggested as a possibility for the surface glow (Ref. H-38). Reaction of NO with ramming O(I)

$$NO + O(I) + (M) \longrightarrow NO_2^* + (M)$$

was suggested. H-14).

The reaction of H and NO, which has not been considered, appears to radiate in the appropriate spectral range, and the intensity appears to increase toward the long wavelengths. Table 3.2-2 shows the results of laboratory measurements (Ref. H-39). Although it was stated that the emission was extremely low, it may be worth further consideration.

Table 3.2-2 Emission for H + NO ⟶ HNO*

Emission, Å	Intensity
6172	Weak
6935	Moderate
7822	Strong
8292	Moderate
8795	Moderate
9500	Weak

From the results of Reference H-26, it has been suggested that any ion recombining with an electron on the surface of Shuttle may yield a reddish glow (Ref. H-40).

Some early studies of ions impinging on metal surfaces showed broadband emissions. This has been observed with 10 keV He^+ and H^+ impinging on copper surfaces (Ref. H-41). A broad unexplained emission centered near 3200Å was observed. The impact of neutral H atoms on the surface gave essentially the same spectrum as for H^+ impact with spectral features of the same relative intensities. The broadband feature cannot be related to the metal because a mixed beam of H^+, H_2^+, and H_3^+ on targets of silver, aluminum, tungsten, and copper impacting at 250 keV gave a similar broadband emission with a maximum near 3300 Å (Ref. H-42). This observation is reminiscent of observations made at Martin Marietta where it was found that 500 to 1000 eV ions of O^+, O_2^+, H_2^+, N_2^+, and He^+ impinging on Chemglaze Z306 all produced emissions that passed through a cutoff filter that only passed the light of wavelengths greater than 6000Å (see Figure 3.2-4). The latter results are in agreement with the observation on AE. Yee and Abreu noted the similar slopes of brightness and atomic oxygen density above 170 km (Figure 3.2-1a) and suggested that the glow was related to atomic oxygen. They claimed "no correspondence with molecular nitrogen was found for either wavelength at any altitude". It is difficult to rationalize this statement when the slopes below the 160 km are considered. At 6563 Å the slopes of brightness and molecular nitrogen appear to be correlated as well as that for brightness with atomic oxygen above 160 km. If the slope of brightness at 7320 Å is drawn through the lower extension of the bar of brightness uncertainty at 140 km (Figure 3.2-1b), a similar correlation can be made for molecular nitrogen and brightness at 7320 Å.

It may be of interest to compare the observed glow intensity on AE with that of observed on Shuttle. From Figure 3.2-2 the total brightness between 4278 Å and 7320 Å can be estimated. For 170-175 km this is calculated to be 9.6 kR (kR = kilo Rayleighs = 10^9 photons/cm^2/sec). On AE the brightness was also determined as a function of altitude. This is depicted in Figure 3.2-1, from which it can be seen that the brightness decreases as the altitude increases. From the constancy of slope for the brightness above 170 km, as well as the comparison of the spectral distribution between Shuttle (Figure 3.2-3) and AE (Figure 3.2-2), it is reasonable to assume a constancy of spectral distribution with altitude for AE. With this assumption the total brightness, B, at any altitude above 170 km between 4278 Å and 7320 Å, may be estimated from Figure 3.2-1 by:

$$\text{Total B at altitude} = \frac{\text{B at 7320 Å at altitude}}{\text{B at 7320 Å at 170 km}} \times \text{total B at 170 km.}$$

Thus at 200 km

$$\text{Total B} = \frac{75R}{180R} \times 9.6 \text{ kR} = 4.0 \text{ kR.}$$

Since AE had a curved surface and the photometer protruded 8 cm above the surface, Dalgarno et. al. (Ref. H-43) suggest a curvature correction of a factor 2 to make it comparable with the flat surface of the stabilizer of Shuttle in the RAM direction and another factor of 2 for the photometer protrusion. Thus, at 200 km the total brightness is 4.0 kR x 4 = 16 kR, to become comparable with a Shuttle flat surface glow. Several points calculated with this correction have been plotted in Figure 3.2-9.

For comparison with the Atmospheric Explorer, results from the analysis of glow photographs on Shuttle flights are also plotted on Figure 3.2-9. The Shuttle flights were STS-3 (240 km), STS-5 (305 km), STS-8 (222 km), and STS-41G (230 and 360 km). For STS-8 and STS-41G, the brightness in R/A are shown in Table 3.2-3 (Ref. H-44 and H-45).

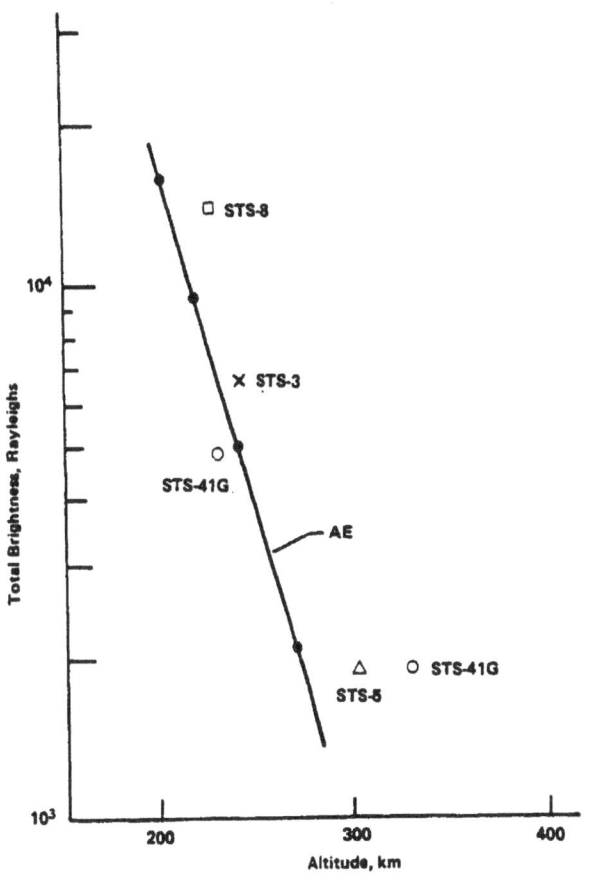

Figure 3.2-9
Brightness of Shuttle Glow Compared with Atmospheric Explorer Glow between 4278Å and 7320Å

Table 3.2-3
Brightness of Glow as Measured on STS-41G and STS-8

Wavelength Å	Brightness R/Å		
	STS-41G 360 km	230 km	STS-8 222 km
5577	<20	≤ 50	150
6300	<35	90	300
7300	<60	≤140	400
7600	<70	≤160	500

Note:
STS-41G—Ref H-44; STS-8—Ref H-45

From Table 3.2-3, the total brightness for STS-8 over the approximately 900 cm of column length across the vertical stabilizer (see Figure 3 of Reference H-14) is estimated to be 665 kR. In the RAM direction the glow peaks about 20 cm from the vertical stabilizer (Ref. H-10). To be comparable to AE, a photometer viewing normal to the RAM direction would see a column length of about 20 cm. The total glow is approximately (20/900) 665 kR = 14.6 kR and this is the point labeled STS-8 in Figure 3.2-9. For STS-41G, using the equality values for 230 km and the listed values for 360 km in Table 3.2-3 the estimated total brightness for 230 km is 4.6 kR and 1.9 kR for 360 km. These are the points labeled STS-41G in Figure 3.2-9. From the figure it appears that for 360 km the values listed in Table 3.2-3 are very much the upper bounds. Much better agreement would be obtained if the actual values were about 10 percent of those listed for the 360 km altitude.

For STS-3 Yee and Dalgarno (Ref. H-10) estimate the Shuttle glow to be about three times the Earth airglow appearing in the background of the photograph of the Shuttle glow. They take as an estimate of the airglow the value 10 kR which is the atomic oxygen airglow at 5577 Å (Ref H-26). This severely underestimates the photographed airglow since it consists of all radiating species and not just atomic oxygen. The airglow has a brightness of 100 kR (Ref H-26), and thus the correct brightness for STS-3 should be about 300 kR. With the correction for column length in the RAM direction the flow is then estimated as (20/900) 300 kR = 6.7 kR which is the point plotted in Figure 3.2-9.

By comparing the glow photograph from STS-3 (240 km) with the glow photograph of STS-5 (305 km) the glow intensity of STS-5 is estimated to be about 1/3.5 times that on STS-3 (Ref. H-14). Both photographs were made with the same camera/film system and with similar velocity vectors of the spacecraft. This estimate for STS-5 is also plotted in Figure 3.2-9.

Considering the available data, Figure 3.2-9 shows reasonably good agreement between Shuttle data and the AE results. The line drawn through the AE points could be used as an estimate of spacecraft brightness as a function of altitude. The curve neglects, however, long term variation in atmospheric density.

Thruster firings on Shuttle create a great deal of observable light, and in addition to this, there is a marked enhancement of the spacecraft ram glow (Ref. H-14). The integrated video signal of the enhanced glows on the engine pods plotted by a chart recorder is depicted in Figure 3.2-10 for STS-3 and STS-5 (Ref. H-14). The decay of glows, beginning at t = 0 are analyzed in Figures 3.2-11 and 3.2-12 where each decay appears to have two time constants. It can be noted that the 20 km difference in altitude between STS-3 and STS-5 results in decay time constants that are 10 times larger for STS-5 than those of STS-3, i.e. for the fast decay the 1/e decay time for STS-5 is 1.2 sec and for STS-3 it is 0.14 sec, and for the slow reaction the times are 6 and 0.6 sec., respectively.

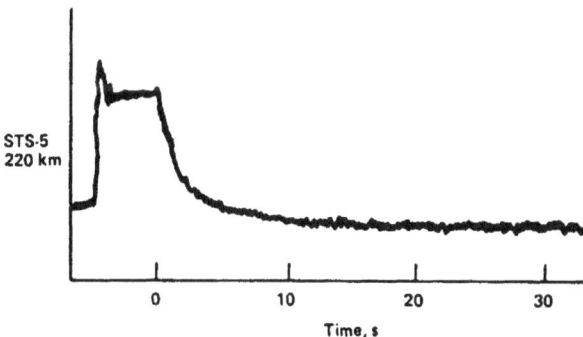

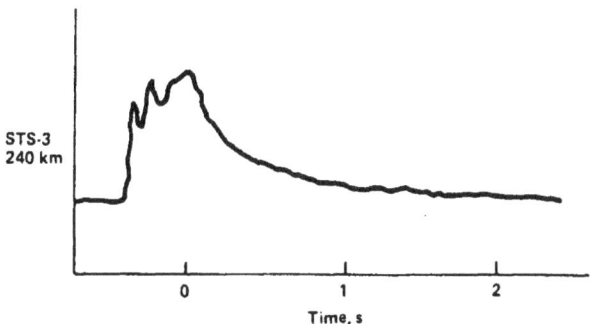

Figure 3.2-10
The Function of the Thruster Glow Intensity on the Engine Pods as a Function of Time after a Thruster Firing (Arbitrary Units) (Ref H-14)

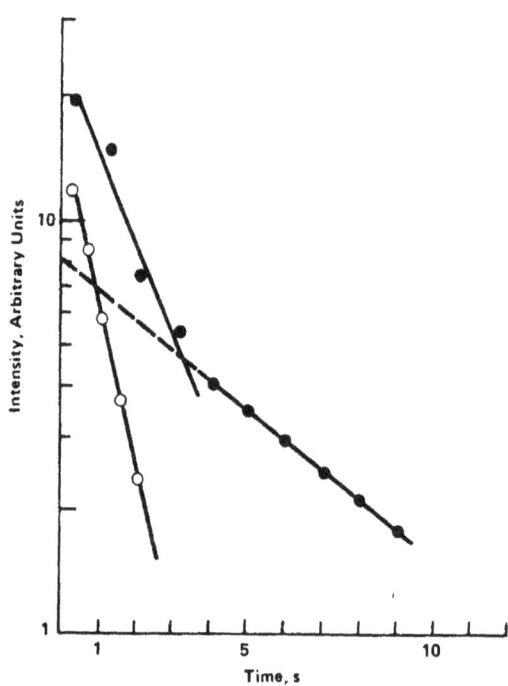

*Figure 3.2-11
Analysis of Decay of STS-5 from Figure 3.2-10
(Solid Points) and Resolution of Data Showing
Fast Reaction (Open Points)*

3.2.6 <u>Key Technical Personnel</u>

	Organization	Phone
K. S. Clifton	MSFC	(205) 544-7725
M. Torr	MSFC	(205) 544-7676
O. K. Garriott	JSC	(713) 483-6581
H. Garrett	JPL	(818) 354-2644
J. H. Yee	Harvard-Smithsonian	(617) 495-5873
S. B. Mende	Lockheed	(415) 858-4082
G. R. Swenson	Lockheed	(415) 858-4097
B. D. Green	PSI	(617) 475-9030
K. Papadopoulos	SAI	(301) 454-6810

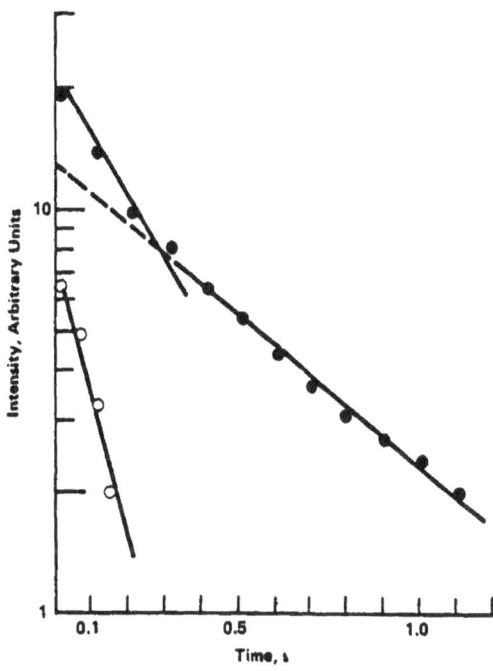

*Figure 3.2-12
Analysis of Decay of STS-3 from Figure 3.2-10
(Solid Points) and Resolution Showing Fast
Reaction (Open Points).*

3.2.7 Glow References

H-1. "Orbital Glow in Dark Concerns NASA", Aviation Week and Space Technology, 116, 14, May 10, 1982.

H-2. Torr, M. R., P. B. Hays, B. C. Kennedy, and J. C. G. Walker, "Intercalibration of Airglow Observations with the Atmospheric Explorer Satellite", Planet Space Sci., 25, 173, 1977.

H-3. Torr, M. R., and D. G. Torr, "The Spectral Assessment of the Spacelab 1/ Shuttle Optical Environment", J. Geophys. Res., submitted 1984.

H-4. Bixler, J., S. Bowyer, J. M. Deharveng, G. Courtes, R. Malina, S. Chakrabarti, and M. Lampton, "The Diffuse For UV Background Observed by the FAUST Telescope on Spacelab One", Space Sciences Laboratory Report, Univ. of Calif., Berkeley.

H-5. Yee, J. H., and V. J. Abreu, "Visible Glow Induced by Spacecraft Environment Interaction", Geophys. Res. Lett., 10, 126, 1983.

H-6. Yee, J. H., and V. J. Abreu, "Optical Contamination on the Atmospheric Explorer-E Satellite", Proceedings of SPIE on Spacecraft Contamination Environment, 338, 120, 1982.

H-7. Abreu, V. J., W. R. Skinner, P. B. Hays, and J. H. Yee, "Optical Effects of Spacecraft-Environment Interaction: Spectrometric Observations by the DE-B Satellite", AIAA-83-2657-CP at Shuttle Environment and Operations Meeting, October 1983.

H-8. Slanger, T. G., "Conjectures on the Origin of the Surface Glow of Space Vehicles", Geophys. Res. Lett., $\underline{10}$, 130, 1983.

H-9. Langhoff, S. R., R. L. Jaffe, J. H. Yee, and A. Dalgarno, "The Surface Glow of the Atmospheric Explorer C and E Satellites", Geophys. Res. Lett., $\underline{10}$, 896, 1983.

H-10. Yee, J. H., and A. Dalgarno, "Radiative Lifetime Analysis of the Shuttle Optical Glow", AIAA-83-2660 at Shuttle Environment and Operations Meeting, October 1983.

H-11. Mende, S. B., O. K. Garriott, and P. M. Banks, "Observations of Optical Emissions on STS-4", Geophys. Res. Lett., $\underline{10}$, 122, 1983.

H-12. Mende, S. B., "Measurement of Vehicle Glow on the Space Emissions Photometric Imaging Experiment", Science, $\underline{225}$ 191, 1984.

H-13. Mende, S. B., G. R. Swenson, and K. S. Clifton, "Atmospheric Emissions Photometric Imaging Experiment," Science, $\underline{225}$, 191, 1984.

H-14. Mende, S. B., and G. R. Swenson, "Vehicle Glow Measurements on the Space Shuttle", in NASA Conference Publication 2391, Second Workshop on Spacecraft Glow, May 1985.

H-15. Mende, S. B., G. R. Swenson, K. S. Clifton, R. Gause, L. Leger, and O. K. Garriott, "Space Vehicle Glow Measurements on STS-41-D", J. Spacecraft Rockets, submitted.

H-16. Dessler, A., "Questions", Minutes of the Physics of Spacecraft Glow Workshop, MSFC, Jan. 19, 1984.

H-17. Mendillo, M., in Space Systems and their Interactions with Earth's Space Environment, H. B. Garrett and C. P. Pike, eds., Progress in Astronautics and Aeronautics, Vol. 71, AIAA 1980.

H-18. Narcisi, R., E. T. Trzcinski, G. Federico, L. Wlodyka, and D. Delorey, "The Gaseous and Plasma Environment Around Space Shuttle", AIAA-83-2659 at Shuttle Environment and Operation Meeting, October 1983.

H-19. Torr, M. R., and D. G. Torr, "A Preliminary Spectroscopic Assessment of Spacelab 1/Shuttle Optical Environment", Minutes of Surface Interactions Subpanel, Shuttle Environment and Experiment Workshop, Aug. 6, 1984.

H-20. Mumma, M. J. and D. E. Jennings, "Planned Investigation of Infrared Emissions Associated with the Induced Spacecraft Glow: A Shuttle Infrared Glow Experiment (SIRGE)", in NASA Conference Publication 2391, Second Workshop on Spacecraft Glow, May 1985.

H-21. Spear, K. A., G. J. Ucker and K. Tobiska, "Survey of Ultraviolet Shuttle Glow", in NASA Conference Publication 2391, Second Workshop on Spacecraft Glow, May 1985.

H-22. Swenson, G. R., and S. B. Mende, "Data Requirements for Verification of Ram Glow Chemistry", in NASA Conference Publication 2391, Second Workshop on Spacecraft Glow, May 1985.

H-23. Torr, M. R., "A Possible Glow Experiment for the EOM-2 Mission", in NASA Conference Publication 2391, Second Workshop on Spacecraft Glow, May 1985.

H-24. Anderson, H. R., "The Shuttle Glow: A Program to Study the Ram-Induced Phenomena", in NASA Conference Publication 2391, Second Workshop on Spacecraft Glow, May 1985.

H-25. Ferguson, D. C., "Laboratory Degradation of Kapton in a Low Energy Oxygen Ion Beam," AIAA-83-2658-CP.

H-26. Bareiss, L. E., H. A. Papazian, and S. K. Anderson, Martin Marietta IRAD, 1984.

H-27. Fontijn, A., C. B. Meyer and H. I. Schiff, "Absolute Quantum Yield Measurements of the NO-O Reaction and Its Use as a Standard for Chemiluminescent Reactions", J. Chem Phys., 40, 64, 1964.

H-28. Paulsen, D. E., W. F. Sheridan and R. E. Huffman, "Thermal and Recombination Emission of NO_2", J. Chem Phys., 53, 647, 1970.

H-29. Schwartz. S. E. and H. S. Johnston, "Kinetics of Nitrogen Dioxide Flourescence", J. Chem Phys, 51, 1286, 1969.

H-30. Swenson, G. R., S. B. Mende, and K. S. Clifton, "Space Shuttle Ram Glow: Implications of NO_2 Recombination Continuum", in NASA Conference Publication 2391, Second Workshop on Spacecraft Glow, May 1985.

H-31. Banks, P. M., P. R. Williams, and W. J. Raitt, "Space Shuttle Glow Observations", Geophys. Res. Lett., 10, 118, 1983.

H-32. Papadopoulos, K., "On the Shuttle Glow", Radio Sci., 19, 571, 1984.

H-33. I. L. Kofsky, and J. L. Barrett, "Optical Emissions Resulting from Plasma Interactions near Windward-directed Spacecraft Surfaces", AIAA-83-2661 at Shuttle Environment and Operations Meeting, October 1983.

H-34. Green, B. D., "Atomic Recombination into Excited Molecular—A Possible Mechanism for Shuttle Glow, Geophys. Res. Lett., 11, 576, 1984.

H-35. Cartwright, D. C., "Vibrational Populations of the Excited States of N_2 Under Auroral Conditions" J. Geophys. Res., 83, 517, 1978.

H-36. Hays, G. N., and H. J. Oskam, "Population of N_2 (B $^3\Pi_g$) by N_2(A $^3\Sigma_u^+$) During Nitrogen Afterglow", J. Chem. Phys., $\underline{59}$, 5107, 1973.

H-37. Mende, S. B., "Experimental Measurement of Shuttle Glow" AIAA-84-0550 at AIAA 2nd Aerospace Sciences Meeting, Reno, January 1984.

H-38. Swenson, G. R., S. B. Mende, and K. S. Clifton, "Ram Vehicle Glow Spectrum: Implications of NO_2 Recombination Continuum", Geophys. Res. Lett., Feb, $\underline{12}$, 97, 1985.

H-39. Cushion, J. K., and J. C. Polanyi, "Infrared Chemiluminescence from Gaseous Reaction Atomic H Plus NO;HNO in Emission", J. Chem. Phys., $\underline{30}$, 317 1959.

H-40. Papazian, H. A., Martin Marietta Internal Memo, July 1984.

H-41. Kerkdijk, C., and E. W. Thomas, "Light Emission Induced by H^+ and He^+ Impact on a Clean Copper Surface", Physica, $\underline{63}$, 577, 1973.

H-42. Chaudri, R. M., M. Y. Khan, and M. M. Chaudri, "Proc. of VIth Int. Conf. on Ionization Phenomena in Gases. Vol. II, 21, 1963.

H-43. Dalgarno, A., J. H. Yee and M. LeCompte, "The Atmospheric Explorer and the Shuttle Glow", in NASA Conference Publication 2391, Second Workshop on Spacecraft Glow, May 1985.

H-44. Kendall, D. J., E. J. Llewellyn, R. L. Gattinger and S. B. Mende, "Orbiter Glow at High Spectral Resolution", AIAA-85-7000, AIAA Shuttle Environment and Operations II Conference, Nov. 1985.

H-45. Kendall, J. J., R. L. Gattinger, E. J. Llewellyn, I. C. McDude, and S. B. Mende, "Orbiter Glow Observations at High Spectral Resolution", in NASA Conference Publication 2591, Second Workshop on Spacecraft Glow, May 1985.

Section 4—Other Environments

SECTION 4

OTHER ORBITAL NATURAL ENVIRONMENTS

The Earth's atmospheric density, temperature and composition are functions of: (1) solar activity; (2) geomagnetic activity; (3) time of day; (4) day of the year; (5) altitude; and (6) latitude.

To the present time Shuttle has flown in the altitude range of 200-400 km, in that portion of the atmosphere known as the ionosphere. The various regions of the atmosphere are depicted in Figure 4-1. The ionosphere exists as a result of ionization by solar radiation and cosmic rays. It extends outward from about 60 km until it merges with the plasma of outerplanetary space. Figure 4-2 (Ref. I-1) shows a schematic view of the upper regions depicted in Figure 4-1.

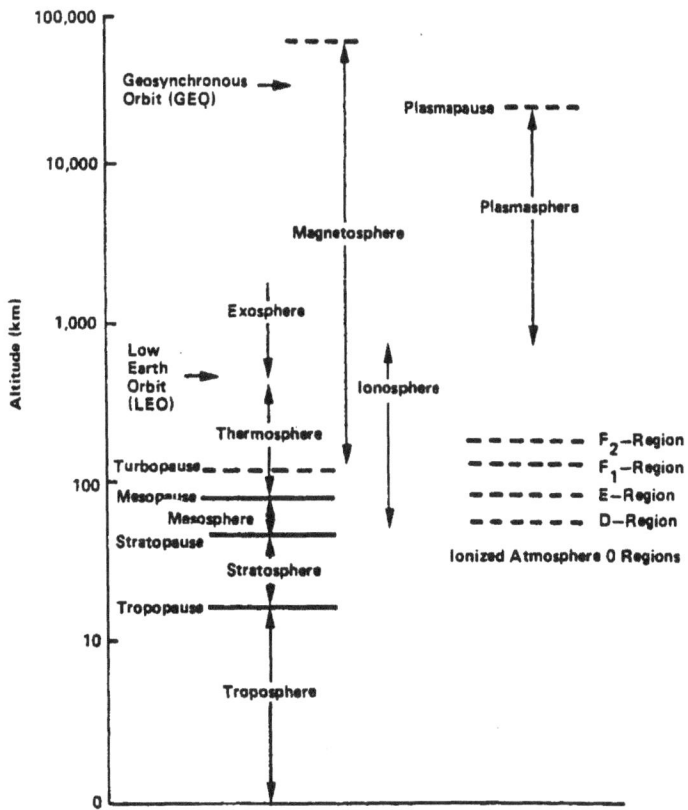

Figure 4-1 Regions of the Earth's Atmosphere

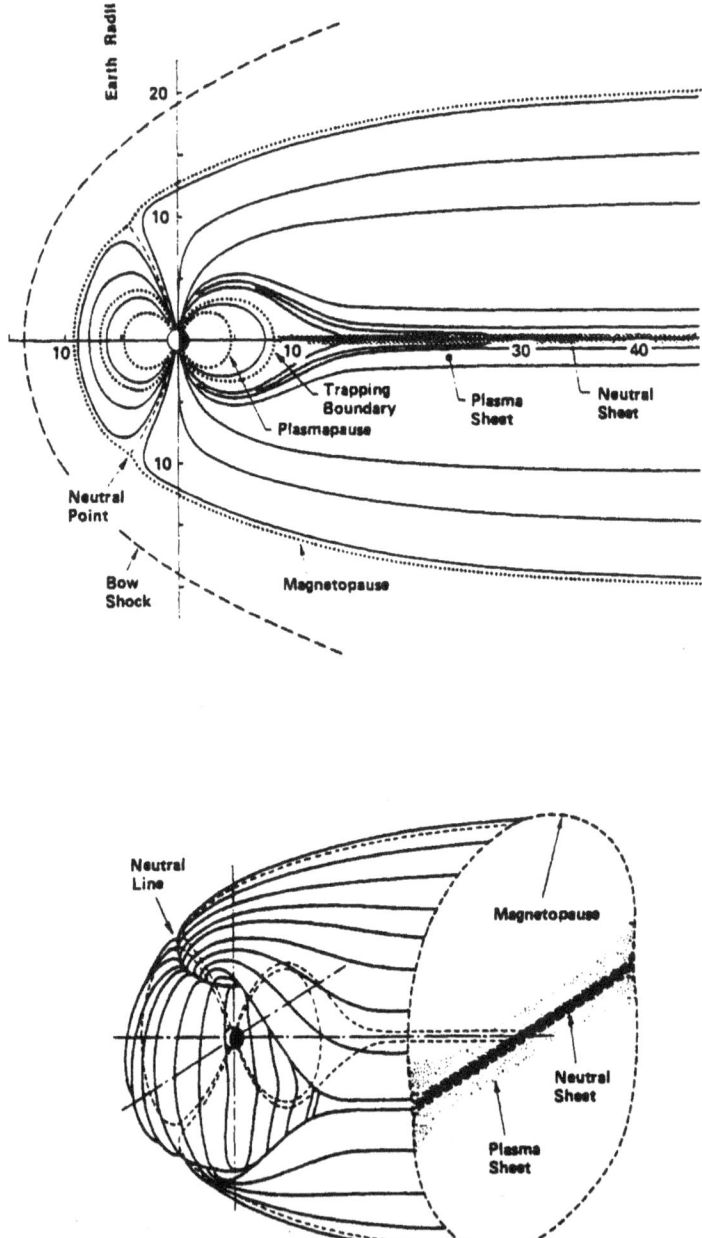

Figure 4-2 General Configuration of the Magnetosphere (Ref 1-1)

The ionosphere is of importance because the interaction of the spacecraft and medium must be taken into account in the design of experiments. These interactions include vehicle drag and heating primarily by the neutral species, the generation of electrical potentials on and about the spacecraft by the ambient plasma and magnetic field interactions, and the formation of wakes and plasma sheaths by passage of the spacecraft through the medium. Therefore in this section, the natural environment of the Shuttle in orbit will be described in terms of: (a) the neutral atomic and molecular species; (b) the plasma, i.e., electrons and charged atomic and molecular species; (c) radiation, both electromagnetic and trapped charge particles,; and (d) the electric and magnetic fields.

4.1 NEUTRAL SPECIES

There are three models described herein: (1) 1976 U. S. Standard Atmosphere (COESA), (Ref. I-2); (2) the Jacchia atmosphere, J71 of 1971 and J77 of 1977 (Ref. I-3 and I-4) and the MSFC/J70 atmosphere derived from the J70 and J71 atmospheres (Ref. I-5 and I-6); and (3) the 1979 mass spectrometer incoherent scatter (MSIS) atmosphere (Ref. I-7 and I-8).

The U. S. Standard Atmosphere (COESA) represents the average composition, density, and temperature based primarily on theoretical solutions to the hydrostatic equation. The J71 and J77 atmospheres are primarily based upon vehicle drag data. The MSFC/J70 atmosphere is based on the J70 atmosphere, but includes modifications from the J71 atmosphere. The MSIS model is semi-theoretical based upon fitting thousands of experimental data points, obtained from flight mass spectrometer and ground based data, to an associated Legendre polynomial expansion of species density variation. It provides detailed information with an estimated accuracy of ± 15 percent over the altitude range 120-800 km. A comparative analysis of these various models has been made (Ref. I-9).

Figure 4.1-1 shows the 1976 U. S. Standard Atmosphere description of the number density of the neutral density.

Where detailed information on density, temperature, and composition is required, the MSIS computer tape programs are available from A. Hedin (Ref. I-7 and I-8) which eliminates the need for a lengthy code input to a computer. The MSIS model inputs geomagnetic (A_p) and solar (F10.7) activity to calculate composition, density, and temperature. The geomagnetic activity is related to the proton flux incident on the Earth's atmosphere. Quiet levels would be represented by a A_p = 4, while very high levels of geomagnetic activity would have A_p on the order of 100. The solar activity is related to the 10.7 cm radiation from hydrogen and is an indication of the extreme ultraviolet (EUV) radiation from the sun. A value of F10.7 = 75 is a low value of EUV corresponding to values near the minimum of the 11 year solar cycle. A value of F10.7 = 200 corresponds to activity near the maximum of the solar cycle.

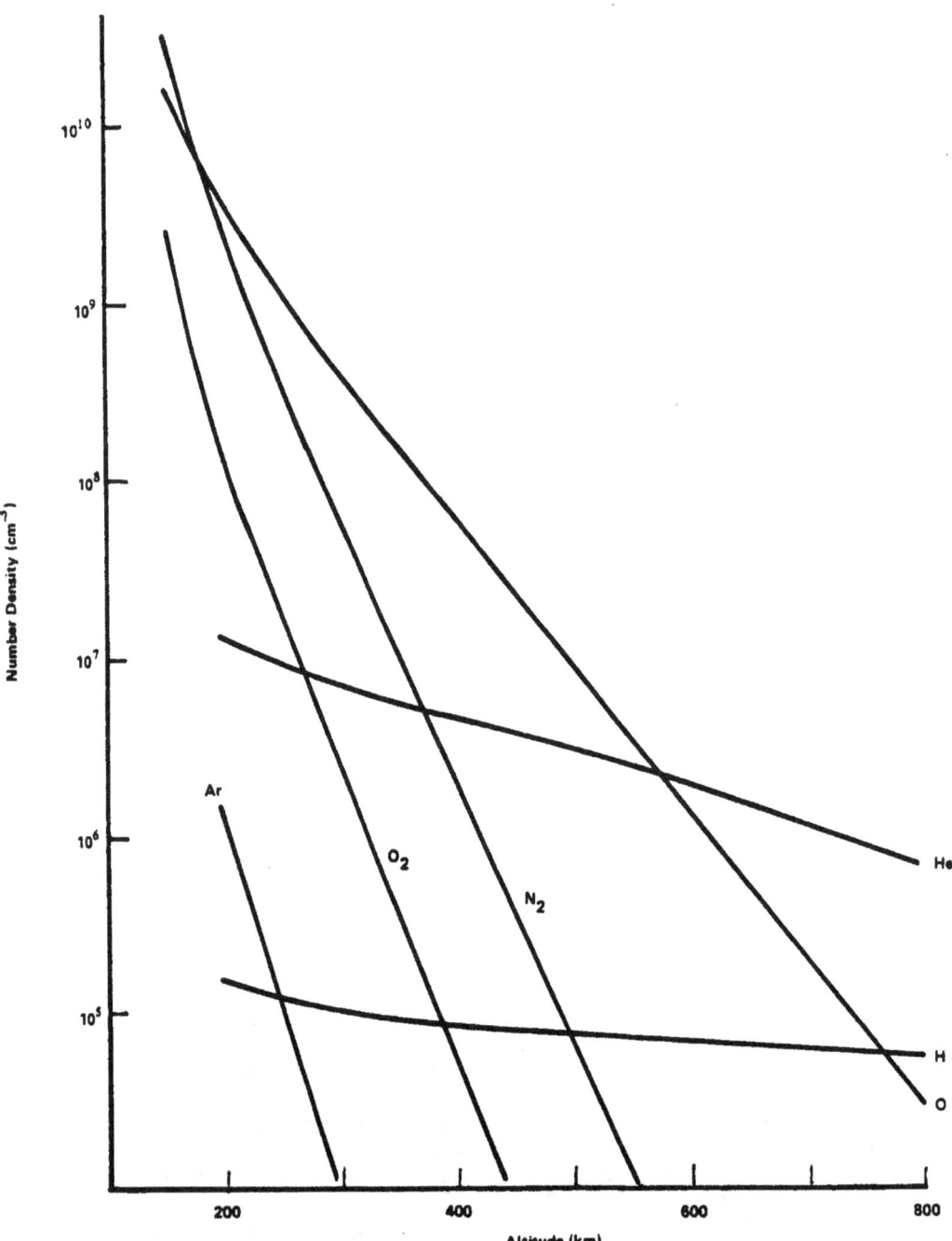

Figure 4.1-1 COESA Model Atmosphere (Ref I-2)

As examples of the MSIS model the temperature, density and composition have been calculated for day 356 and 0° latitude for the two local times: 1500 hours and 0100 hours. These times represent the maximum and minimum density, respectively, in the diurnal variation while day 356 represents the average of density variation throughout the year and can account for any day of the year to within ± 50 percent. Figures 4.1-2, 4.1-3, 4.1-4, and 4.1-5 show the temperature, mass density, and atomic oxygen calculated by the MSIS model as a function of geomagnetic and solar activity on day 356 at 0° latitude. The COESA results are also depicted for comparison with the MSIS model. Results similar to Figure 4.1-5 apply to the other species i.e., hydrogen, argon, helium, oxygen and nitrogen molecules.

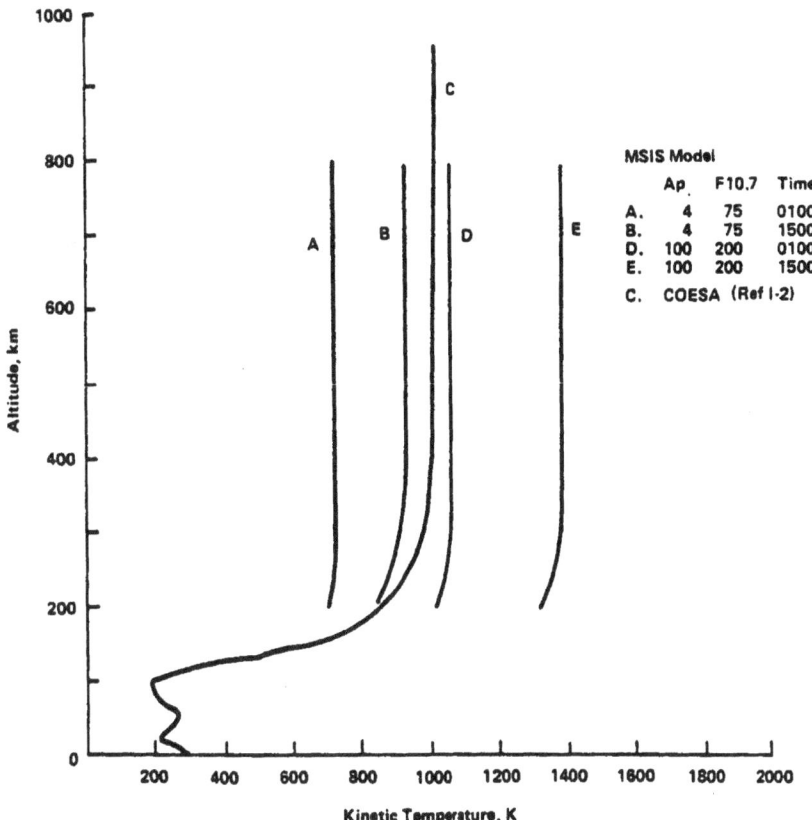

Figure 4.1-2
Temperature vs Altitude: A, B, D, E, on Day 356 at 0° Latitude (MSIS)

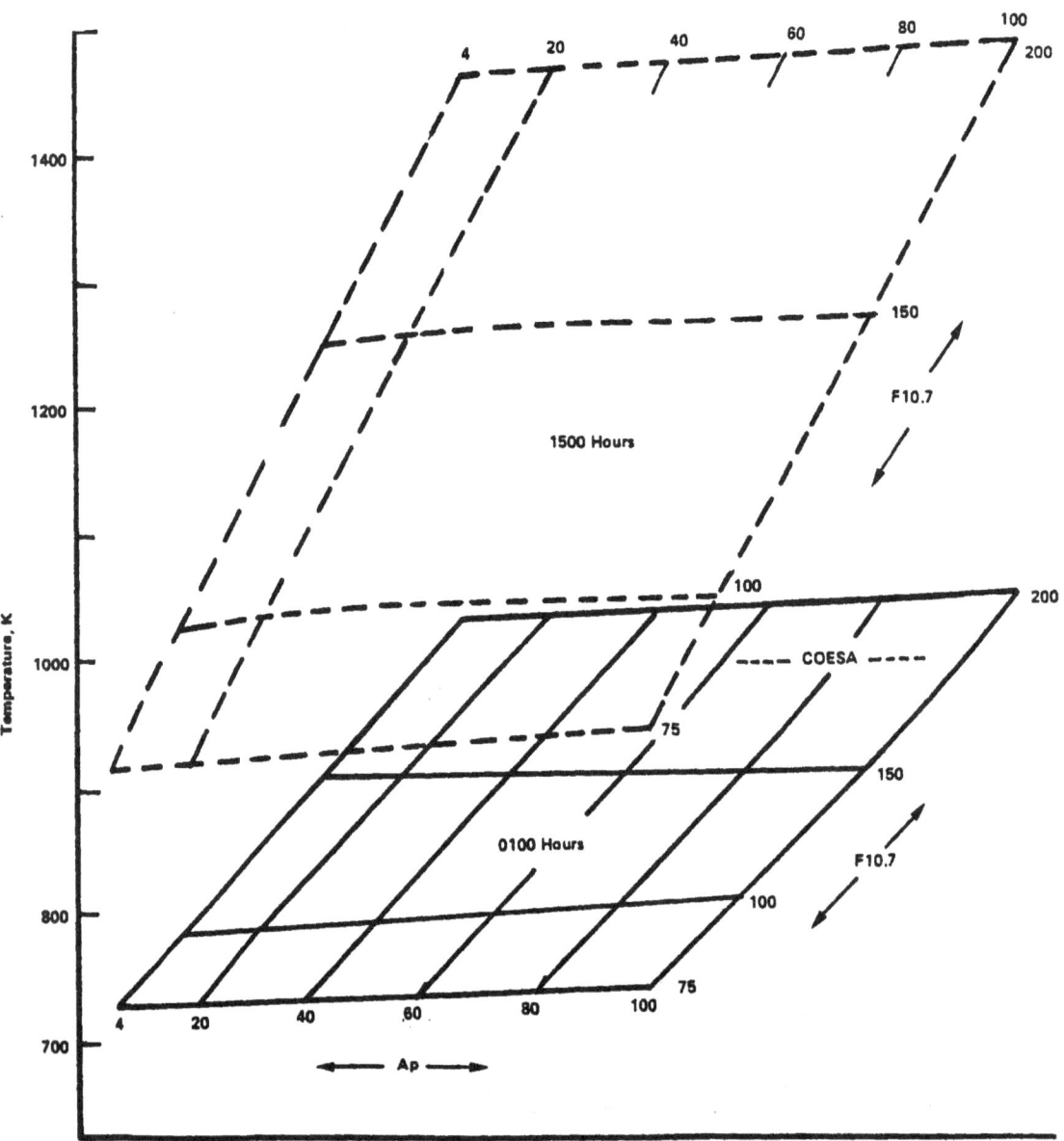

Figure 4.1-3 Temperature above 400 km on Day 356 at 0° Latitude

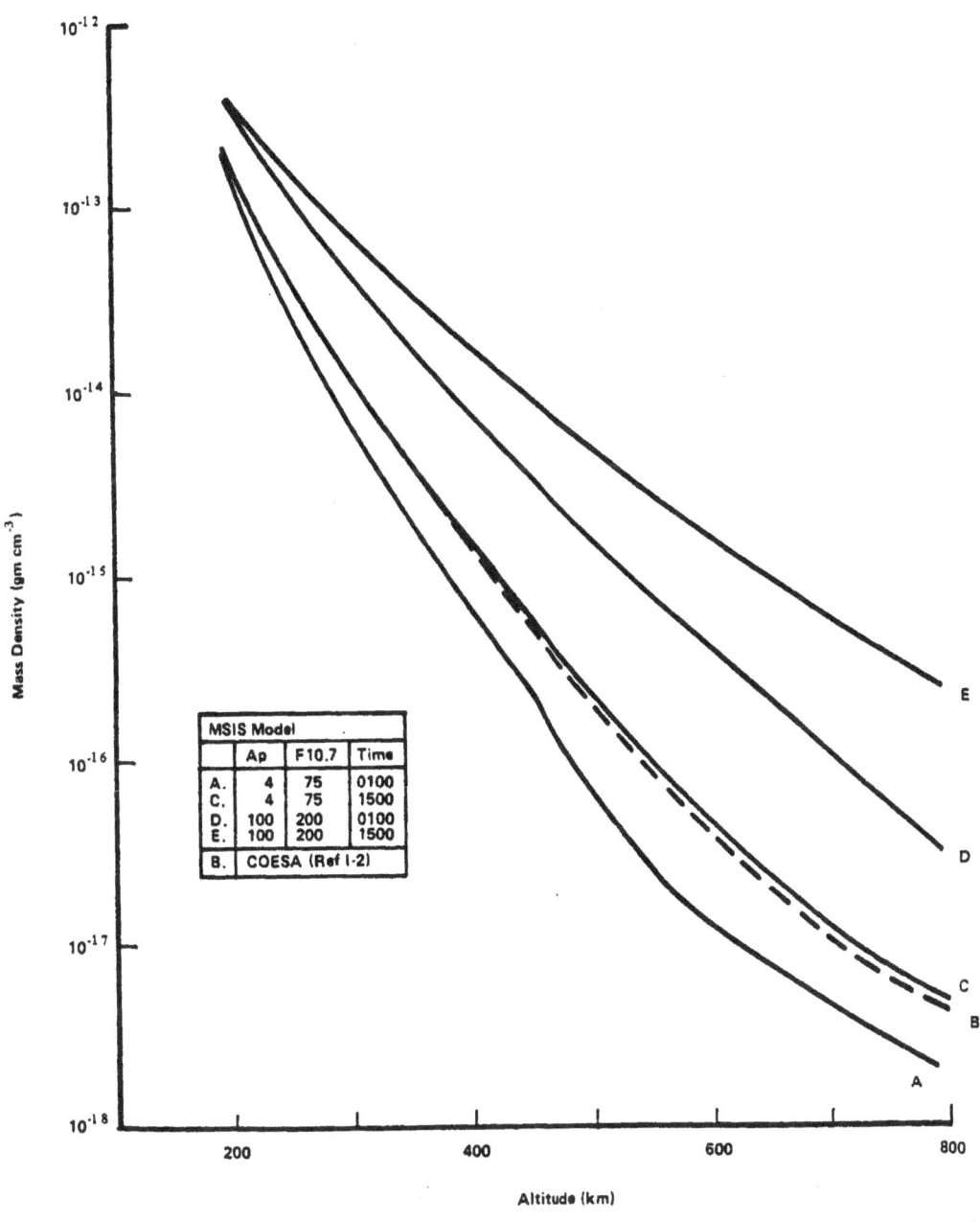

Figure 4.1-4
Mass Density vs Altitude: A, C, D, E on Day 356 at 0° Latitude (MSIS)

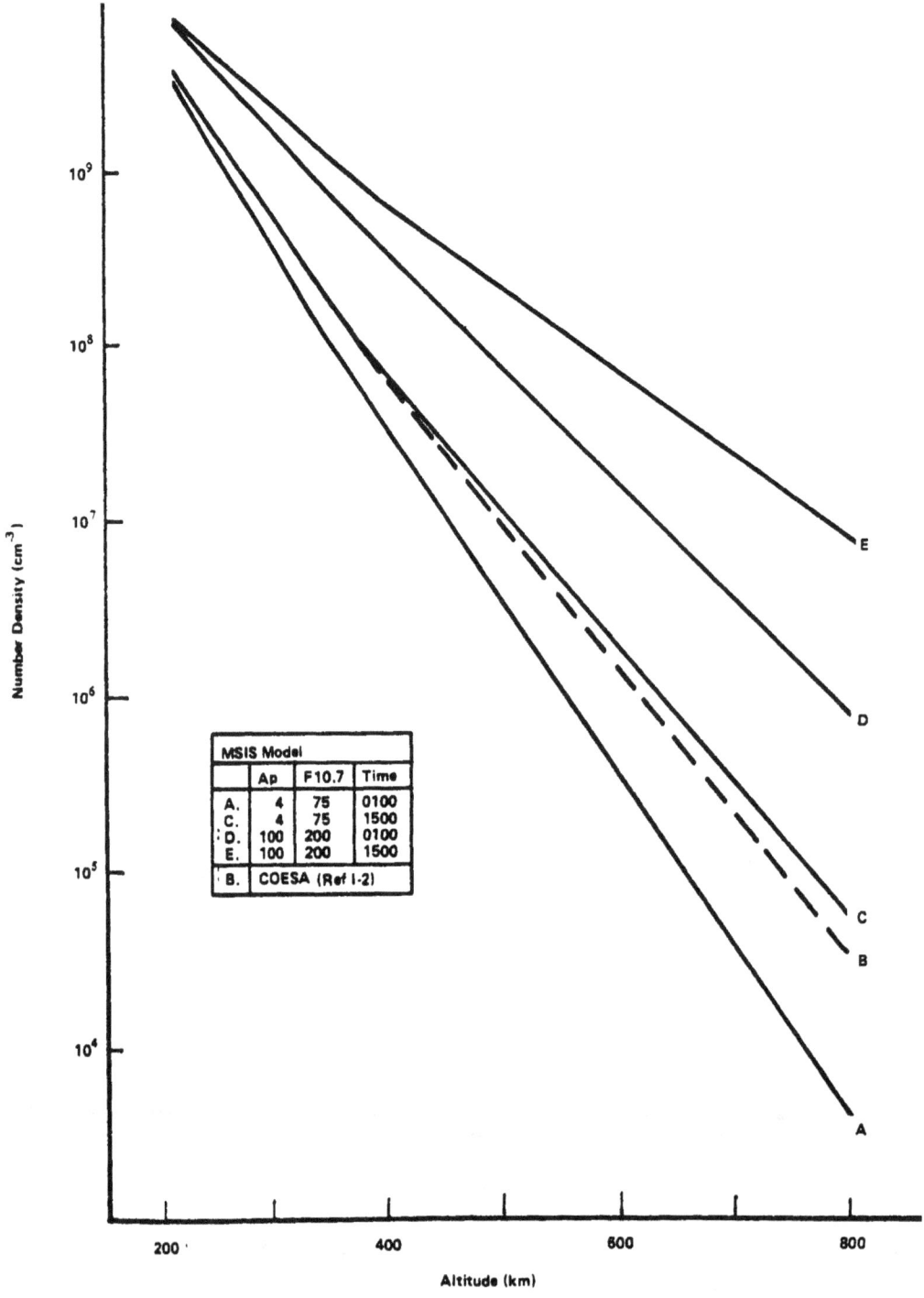

Figure 4.1-5
Atomic Oxygen vs Altitude: A, C, D, E on Day 356 at 0° Latitude

4.2 THE PLASMA

The thermalized charged particles (i.e., atomic and molecular ions and electrons) constitute the plasma. The energetic particles (trapped species) will be discussed in the following section. During the daytime, the lower ionosphere features are identified by three regions. In order of increasing altitude, they are the D, E, and F regions (See Figure 4-1) of which the F region is important for Shuttle. It ranges from about 140 to 1000 km. During the daytime, it has two divisions, designated as F_1 and F_2. The F_1 region is associated with ion production in the vicinity of 150 km and disappears at night as the electron density decreases above the E region. The F_2 region is usually within the altitude range of 200 to 400 km (note that this is the range of present Shuttle flights). It is associated with the peak in the electron density distribution which varies with the time of day, solar cycle, and latitude. The positive ion population is dominated by O^+ ions as can be noted from the ionic densities depicted in Figure 4.2-1. An indication of the electron density distribution is also depicted in Figure 4.2-2. The particle temperatures (and velocities) in the F_2 region are depicted in Figure 4.2-3. The small change of the ionic parameters between 200 and 400 km contrasts sharply with the change of the electron parameters. Figure 4.2-4 shows the effects for electrons of geographical variations in the anomalies of the Earth's magnetic field (Ref. I-10).

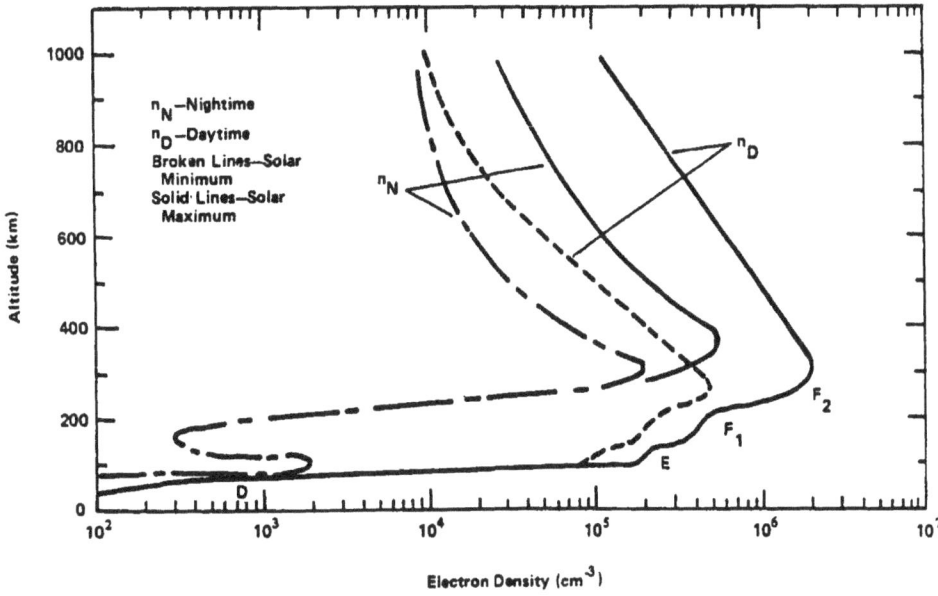

Figure 4.2-2 Ionospheric Electron Concentration

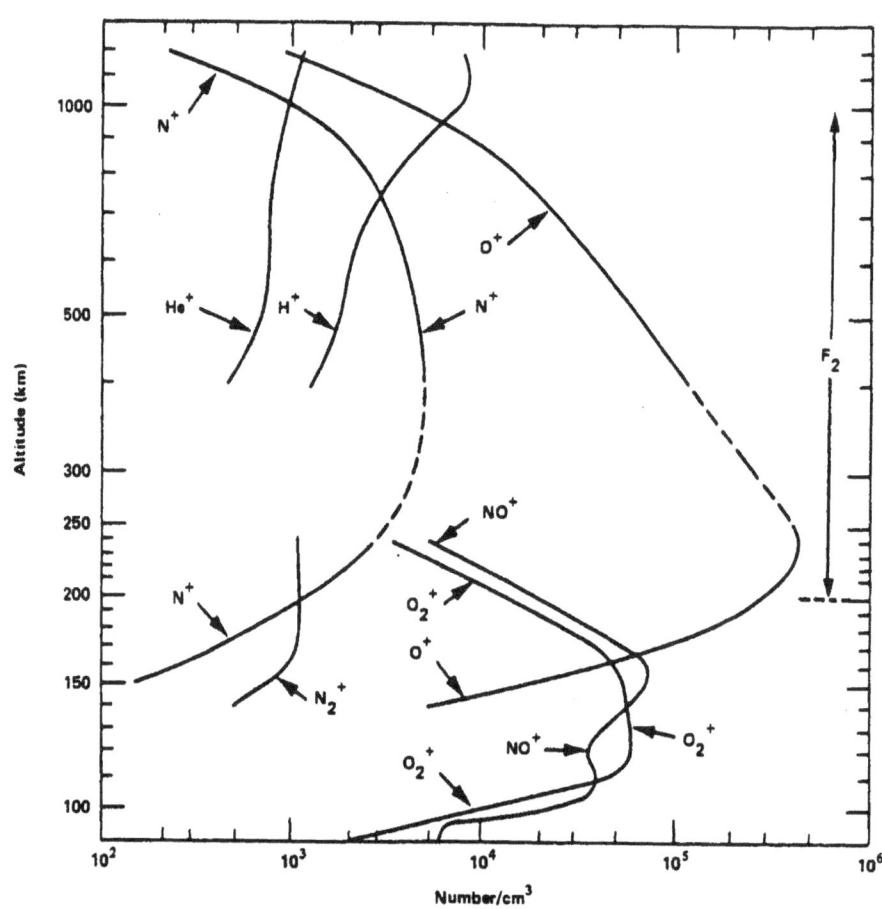

Figure 4.2-1 Ionic Composition of Solar Minimum Daytime Winter Ionosphere

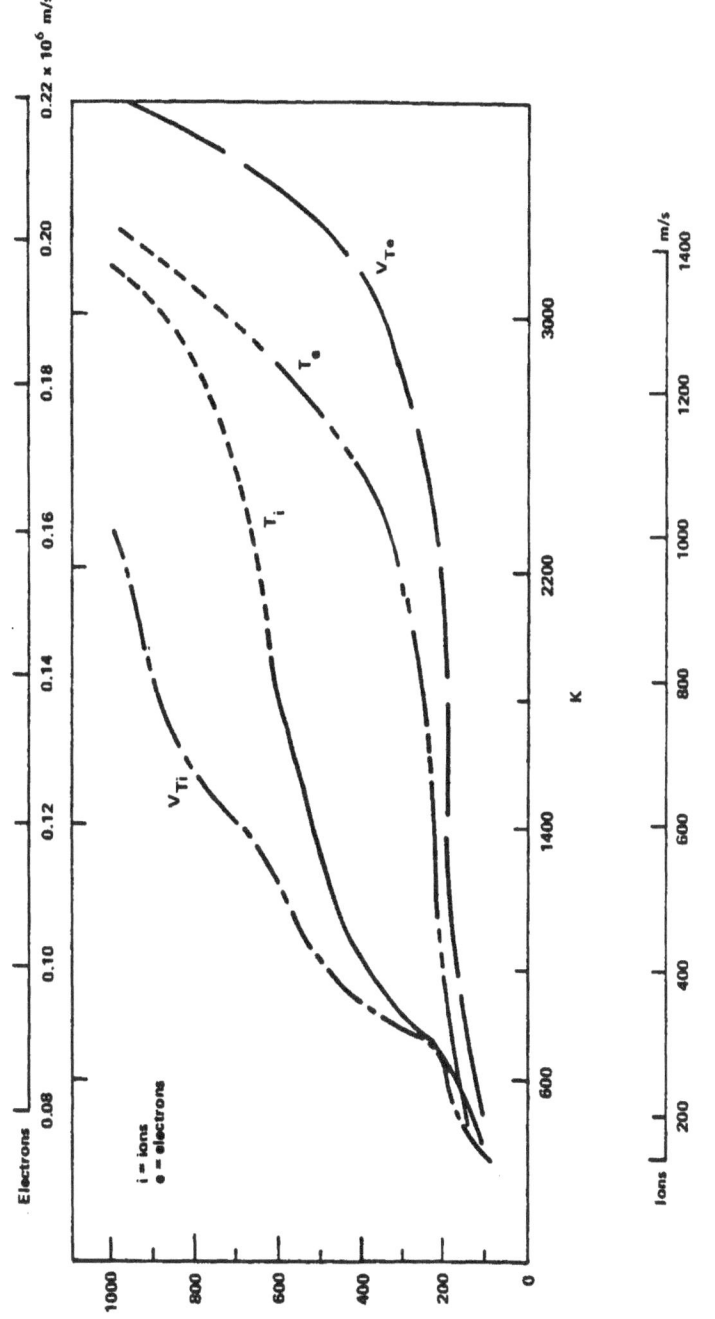

Figure 4.2-3 Particle Temperature and Velocity

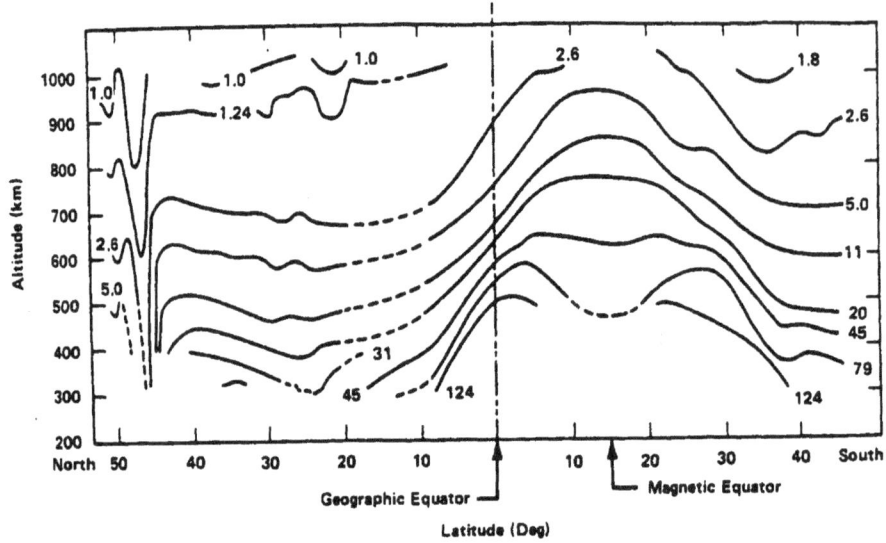

Figure 4.2-4
Daytime Contours of Equal Electron Density in the Orbital Plane of an Alouette Satellite. Density Units of 10^4 Electrons cm^{-3}

4.3 RADIATION

4.3.1 Electromagnetic

The sun is a source of electromagnetic radiation subtending ~ 32.0 min. of arc (0.009931 radians) at 12 A.U (A.U = mean distance from the sun of the earth's orbit). At zero air mass (outside the atmosphere of the earth) the total irradiance is 1396 ± 27 W/m^2 and is essentially constant (Ref. I-11). The seasonal variation amounts to -3.27 percent at aphelion and +3.42 percent at perihelion. The time variation in the far UV, x-ray, and radio-frequency region does not contribute any significant amount to the total irradiance. Figure 4.3-1 shows the spectral distribution at zero air mass as well as that at sea level (Ref. I-12). Table 4.3-1 shows the spectral distribution over various wavelength intervals (Ref. I-11).

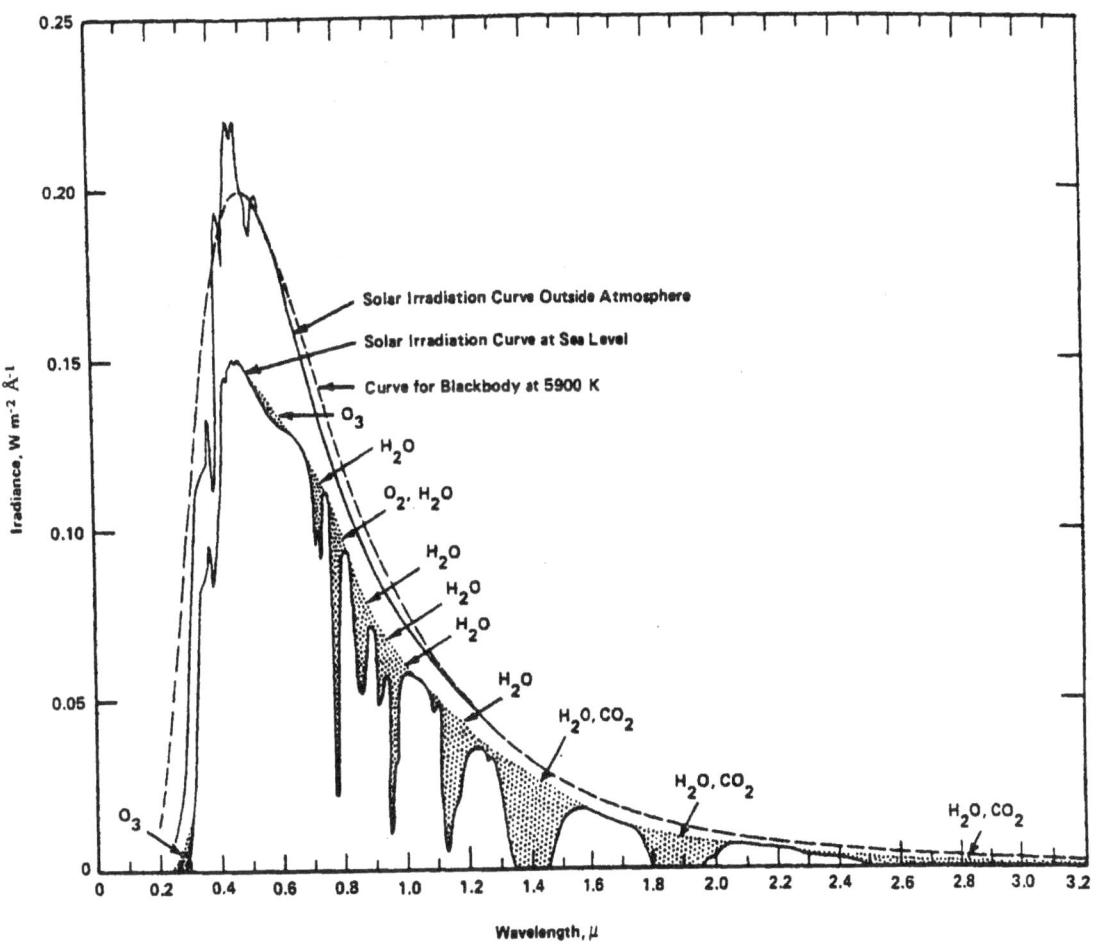

Figure 4.3-1
Spectral Distribution Curves Related to the Sun; Shade Areas Indicate Absorption, at Sea Level, Due to the Atmospheric Constituents Shown

Table 4.3-1 Spectral Distribution of Solar Radiation

Δλ nm	W/m²	Total %	Δλ nm	W/m²	Total %
0-225	0.41	0.03	750- 800	63.56	58.70
225-250	1.40	0.13	800- 850	56.65	62.76
250-275	4.20	0.43	850- 900	50.36	66.36
275-300	11.17	1.23	900- 950	44.72	69.56
300-325	19.10	2.60	950-1000	39.71	72.40
325-350	28.32	4.63	1000-1050	35.07	74.91
350-375	30.87	6.83	1050-1100	31.83	77.18
375-400	30.54	9.02	1100-1500	156.95	88.42
400-425	46.93	12.38	1500-2000	80.90	94.22
425-450	48.00	15.82	2000-2500	35.07	96.73
450-475	54.12	19.70	2500-3000	17.45	97.98
475-500	51.77	23.41	3000-3500	9.62	98.67
500-525	48.50	26.88	3500-4000	5.68	99.08
525-550	49.15	30.40	4000-4500	3.72	99.34
550-575	47.91	33.83	4500-5000	2.28	99.50
575-600	47.44	37.23	5000-6000	2.79	99.70
600-650	86.49	43.42	6000-7000	1.47	99.81
650-700	78.78	49.06	7000-∞	2.67	100.00
700-750	71.02	54.15			

Solar x-rays from the corona are most intense in the vicinity of flares, plages, and sun-spots. The intensity depends on general solar conditions, and in the absence of flares, the total emission below about 50 Å varies from average values of 0.13 to 1.0 ergs/cm²s over a solar cycle. At solar minimum, with a quiet sun, the short wavelength limit is about 10 Å while at solar maximum, with a quiet sun, it extends downward to about 6 Å. Figure 4.3-2 shows emissions in the region 260 to 1300 Å in 1961 which was near the maximum of the moving 11 year mean of sunspot numbers (Ref. I-12). Figure 4.3-3 shows emission associated with a large solar flare in 1969 extending down to 1 Å (Ref. I-13).

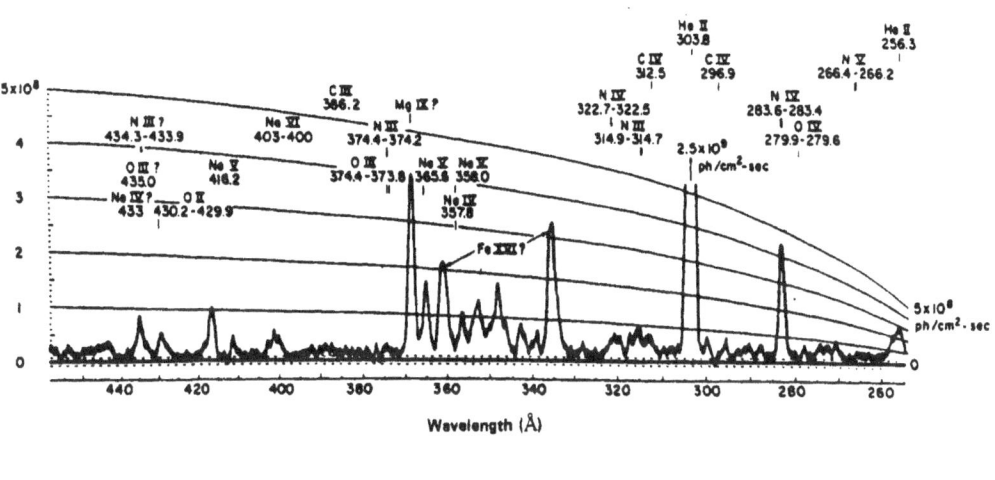
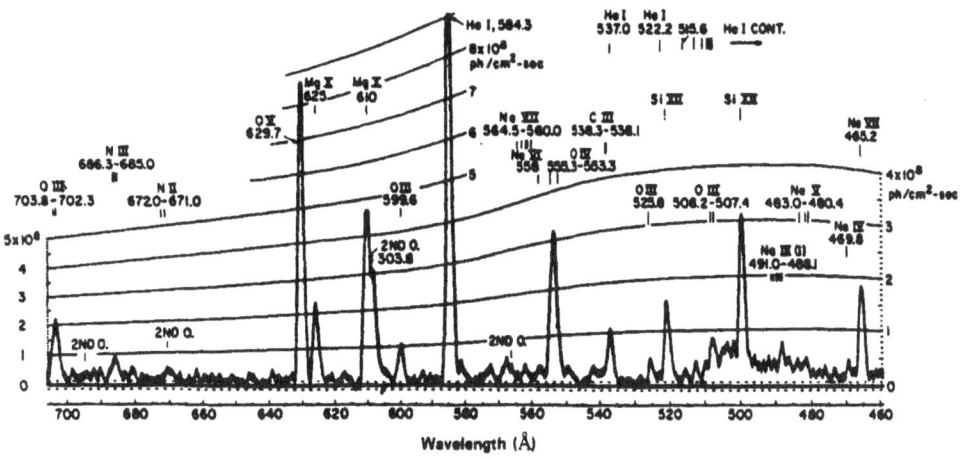

*Figure 4.3-2
Solar spectrum at 225-km altitude averaged from three scans with a grazing incidence monochromator and a photomultiplier-type detector on 23 August 1962, White Sands, NM, 1004 h MST*

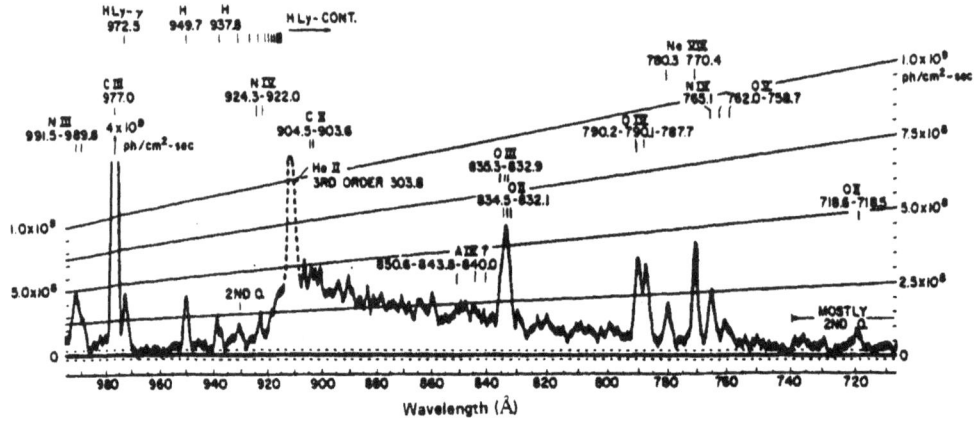
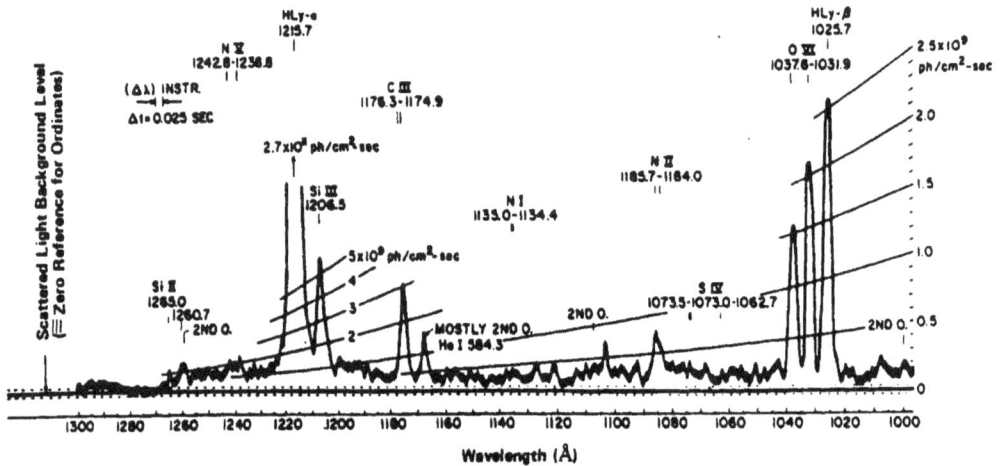

Figure 4.3-2 (concl)

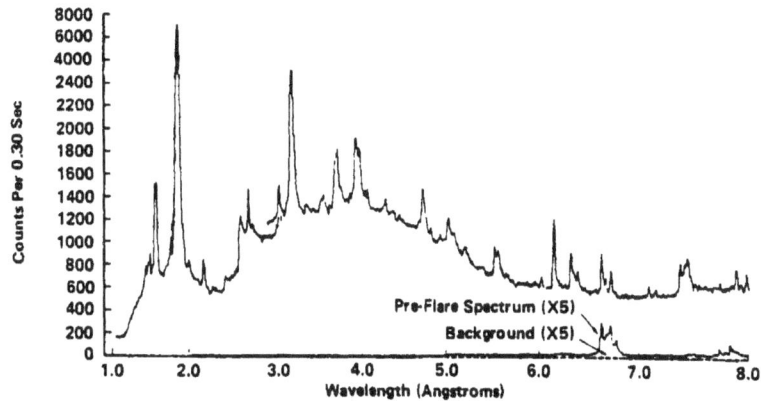

Figure 4.3-3
X-ray spectrum associated with a large solar flare on Feb. 27, 1969

4.3.2 <u>Charged Particles</u>

The low energy charged particles were discussed in Section 4.2. Here the energetic particles in the environment will be considered. These arise from galactic cosmic rays from outside the solar system, solar cosmic rays from bursts (solar flares, etc.) from the sun, and the trapped radiation of the Van Allen belts.

The galactic cosmic rays consist of approximately 85 percent protons, 14 percent alpha particles and about 1 percent of heavier nuclei (Li to Fe). The galactic proton flux amounts to about 4 protons/cm^{2}s, independent of energy in the 10 to 100 MeV energy range. Figure 4.3-4 compares solar and galactic energy spectra (Ref. I-1).

Solar cosmic rays consist of protons, alphas, and electrons of energies generally lower than galactic cosmic rays. Below 1 GeV kinetic energy (1GeV = 10^9eV) and down to about 1 MeV the integral flux of solar protons is about 10^5 greater than the galactic particles. Above 1 GeV there are fewer solar protons than galactic protons. The spectral representation of the proton integrated flux can be given by (Ref. I-14).

$$J(R) = 1.5 \times 10^{11} e^{(-R/88)} \qquad 4.3-1$$

where R is the magnetic rigidity, R = p/Z = momentum/unit charge, in Mv (million volts) and J is /cm^2. Another model (Ref. I-15) for the solar cosmic ray spectra is given in Table 4.3-2.

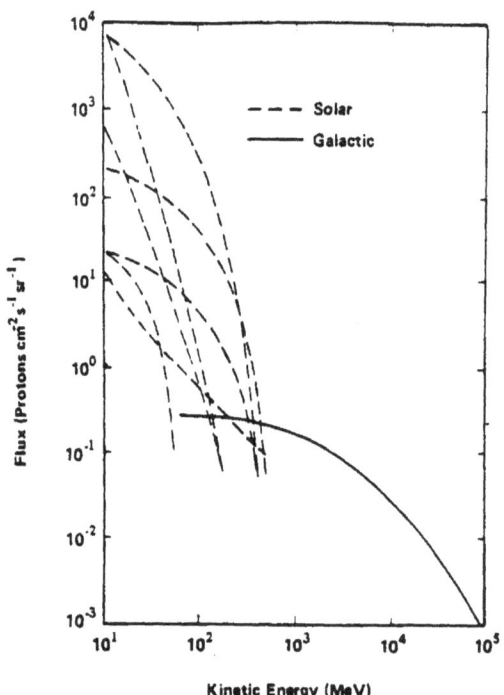

Figure 4.3-4
Energy Spectra of Protons from Several Moderate-Size Solar Events Compared with the Galactic Cosmic Ray Spectrum at Solar Minimum (Ref I-1)

Table 4.3-2 Model Solar Cosmic Ray Spectra

Protons:	$N_p (>T) =$	$7.25 \times 10^{11} \; T^{-1.2}$	; 1 MeV $\leq T \leq$ 10 MeV
		$3.54 \times 10^{11} \; e^{-P(T)/67}$	; 10 MeV $\leq T \leq$ 30 MeV
		$2.64 \times 10^{11} \; e^{-P(T)/73}$	; $T \geq$ 30 MeV
Alphas:	$N_\alpha (>T)$	$N_p (>T)$	; $T <$ 30 MeV
		$7.07 \times 10^{12} \; T^{-2.14}$	; $T \geq$ 30 MeV.

T = kinetic energy = $E - m_o c^2$
P = RZ = momentum

These cosmic rays are shielded from particular latitudes by the Earth's magnetic field. In the dipole approximation, the minimum rigidity that can reach a particular location depends on (Ref. I-16):

$$R_{min} = \frac{60\cos^4\theta}{L^2[1 + (1 - \cos^3\theta\cos\gamma)^{1/2}]^2} \text{ GV} \qquad 4.3-2$$

where

L = radial distance in Earth radii;
θ = latitude; and
γ = half angle of allowed cone of arrival direction about the normal to the meridian plane.

Figure 4.3-5 shows a geomagnetic exposure map for solar protons (for a spacecraft in circular orbits) assuming all solar protons are excluded from the magnetosphere at latitudes less than 63.4° (r < 5) and all solar protons have free access above that latitude. The region under the zero percent curve encompasses orbits that are completely inaccessible. Below 1000 km the inaccessible region reaches up to an orbit inclination of about 50°. Thus only orbits with tilts greater than 50° will encounter solar protons from 0 to about 32 percent of their lifetime.

The energetic electrons and protons trapped in the Van Allen radiation belts will produce the majority of the radiation damage in an orbit as the Shuttle moves through the South Atlantic Anomaly. Figure 4.3-6 shows the anomaly and the path of several orbits, and Figure 4.3-7 shows the flux history over a 24 hour period in a 593 km orbit at 28.8° inclination (Ref. I-17). The results of Figure 4.3-7 were calculated by the Vette model (Ref. I-18 and I-19) which is the generally used model for such calculations. Passage through the anomaly last about 15 minutes at the lower altitudes. Figures 4.3-8 and 4.3-9 show the trapped fluxes for electrons and protons at 28.8° as a function of energy and altitude.

The motions of the particles are depicted in Figure 4.3-10. In the "guiding center" approximation the motion is separable into three components (Ref. I-20). The first component is a circular motion perpendicular to the magnetic field lines with the local cyclotron period T_1, and cyclotron radius R_c where (in gaussian units)

$$T_1 = \frac{2\pi mC}{eB} \qquad R_c = \frac{v_\perp mC}{eB} \qquad 4.3-3$$

where m, e, and $v_\perp$ are the particle mass, charge, and velocity component perpendicular to the field lines.

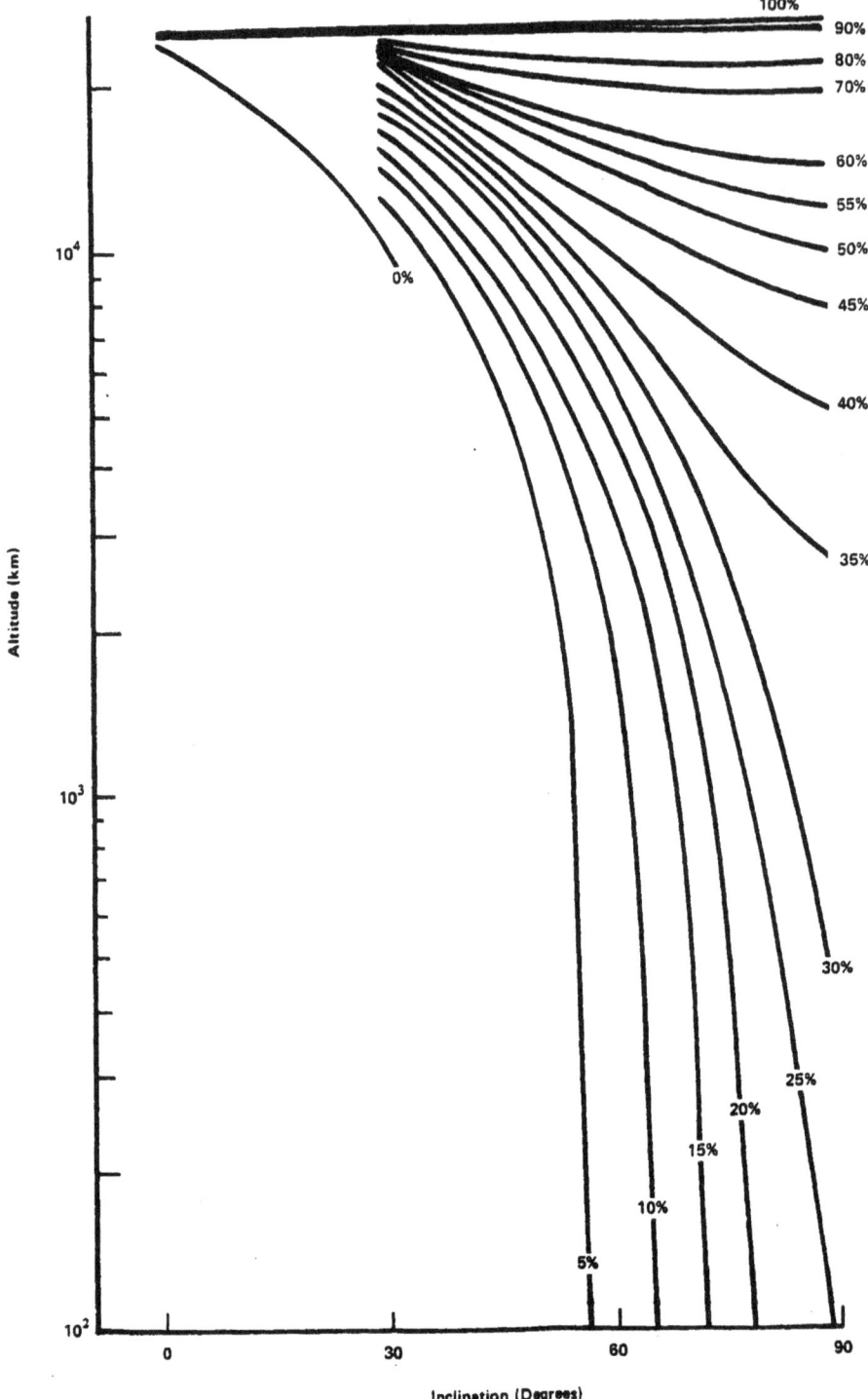

Figure 4.3-5
Percentage of Interplanetary Fluence Intercepted by Spacecraft in Circular Geocentric Orbits as a Function of Altitude and Inclination

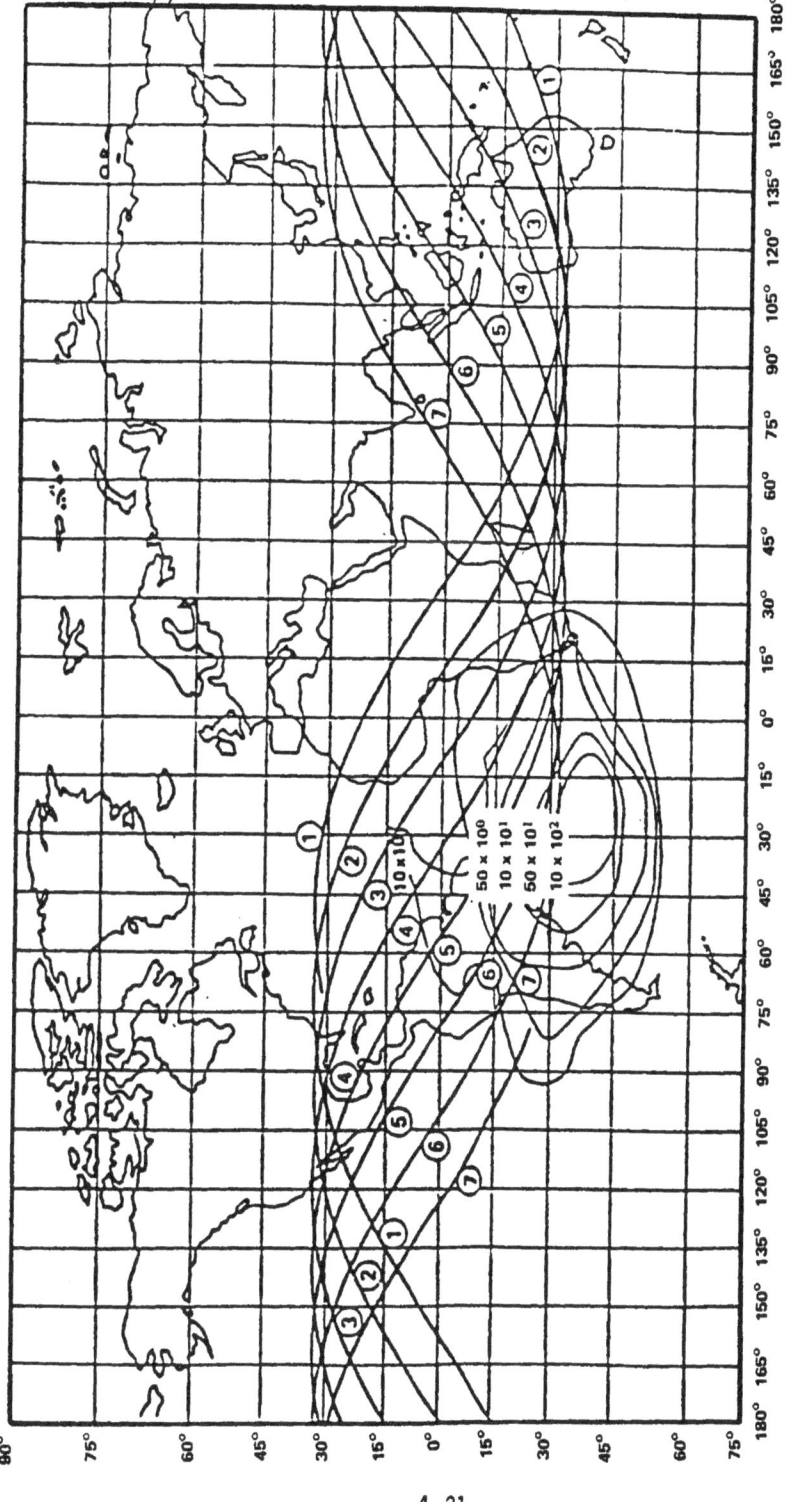

Figure 4.3-6 Proton Flux Densities at an Altitude of 296 km

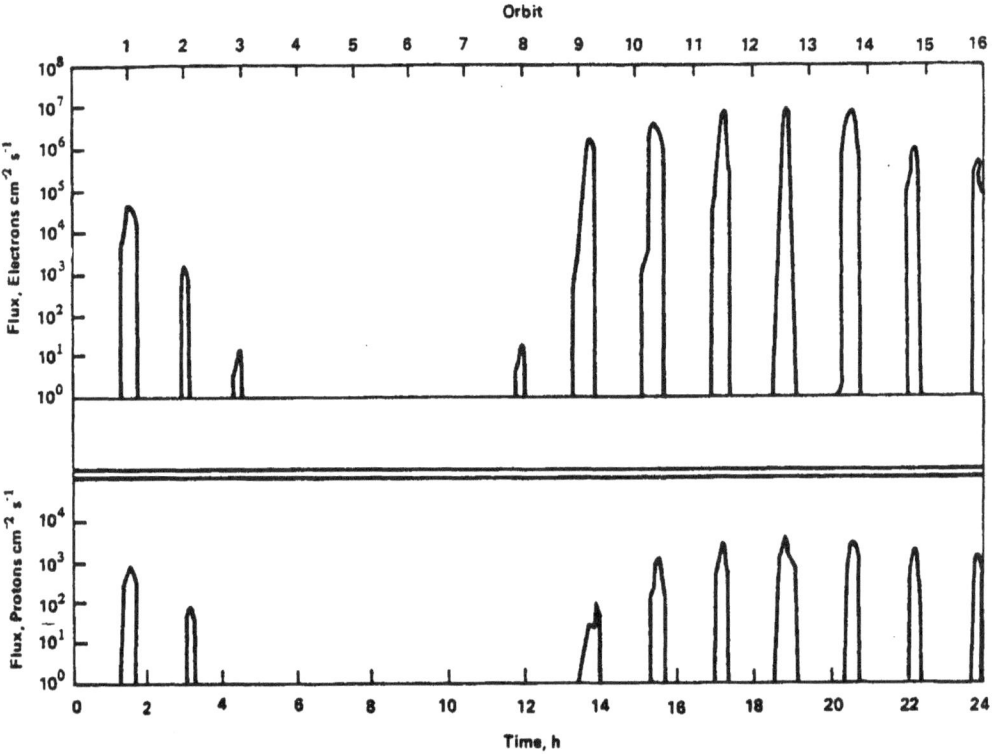

Figure 4.3-7 Flux History over 24-Hour Period (539-km Circular Orbit at 28.8° Inclination)

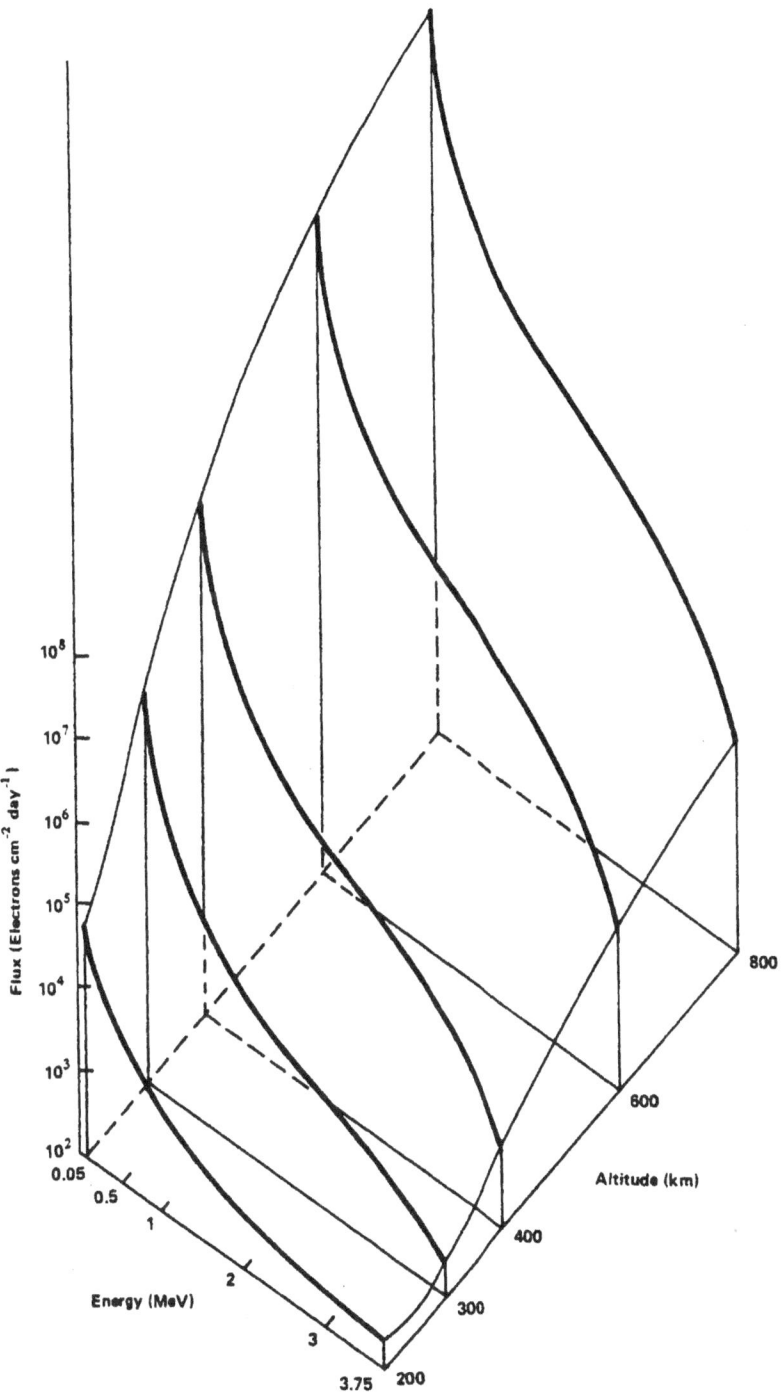

Figure 4.3-8
Trapped Electron Omnidirectional Integral Fluxes above Given Energies at 28.5° Inclinations

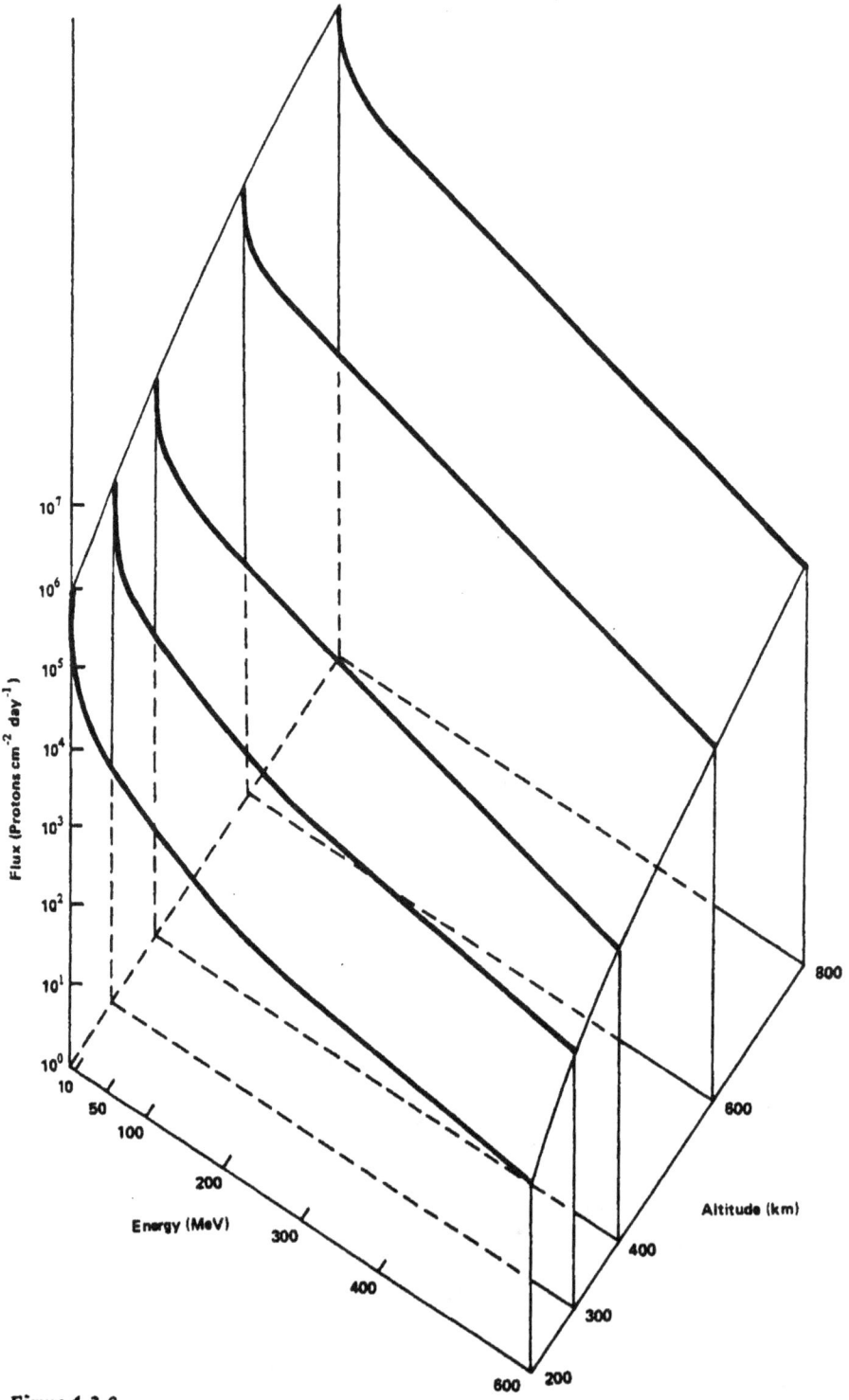

Figure 4.3-9
Trapped Proton Omnidirectional Integral Fluxes above Given Energies at 28.5° Inclination

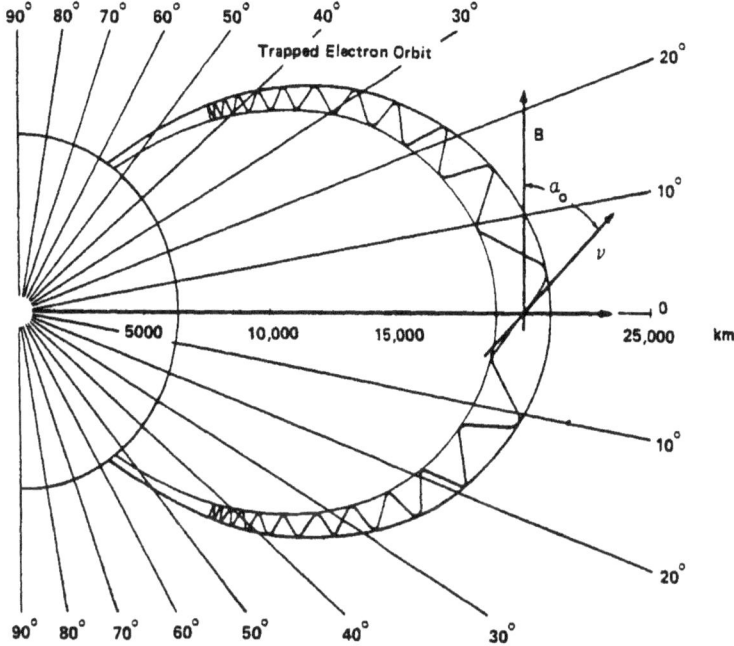

Figure 4.3-10
Schematic representation of adiabatic charged particle motion

The second component is a motion along the field lines in the direction of increasing flux density to a point at which the particle is reflected. This is the mirror point, and it then returns to the other mirror point, called the conjugate point, in the opposite hemisphere. This oscillatory motion has a period T_2, substantially longer than T_1.

Since the magnetic field is static, the total energy E of the particle is conserved and the total flux through the circular orbit is constant. This "first adiabatic invariant" is

$$\frac{1/2 m v_\perp^2}{B} = \frac{1/2 m v^2 \sin^2 \alpha}{B} = \text{constant (non-relativistic particles)} \qquad 4.3\text{-}4$$

and

$$\frac{p^2 \sin^2 \alpha}{B} = \text{constant (relativistic particles)} \qquad 4.3\text{-}5$$

where p is the momentum of the particle and α is the angle between the velocity and field. It then follows that

$$\frac{\sin^2 \alpha_1}{B_1} = \frac{\sin^2 \alpha_2}{B_2} \qquad 4.3-6$$

where the subscripts refer to any two points along the particle trajectory. If the points chosen are at the equator and mirror point, it follows that

$$\frac{B_m}{B_0} = \frac{1}{\sin^2 \alpha_0} \qquad 4.3-7$$

where B_m is the field at the mirror point and B_0 and α_0 are the values of the equator. Particles which have values of α_0 so small that B_m will be at an altitude near 100 km will be removed by atmospheric scattering.

A second "adiabatic invariant" associated with the motion between the mirror points is

$$J = \oint m v_{\parallel} \, ds \qquad 4.3-8$$

where $v_{\parallel}$ is the velocity component along the field lines and the integral is along the line over a complete oscillation between the mirror points.

The third motion is a slow drift in longitude with a period T_3 where electrons drift eastward and protons westward. The particle transfers slowly from one flux tube to another until it returns to its original flux tube. This process generates what is known as a magnetic shell (surrounding the earth and open at both ends). The "third invariant" of motion requires the total number of flux lines passing through this shell to be constant. This statement is trivial for a static field and is important only for (slowly varying) time-dependent fields.

Three periods for typical trapped particles are given in Table 4.3-3.

Protection against radiation damage can be made with an absorber. Figure 4.3-11 shows the total dose rate (protons, electrons, Bremsstrahlung and cosmic rays) behind a spherical aluminum shell for various thicknesses of aluminum as a function of altitude for 28.5° inclination. It can be noted that at present Shuttle altitudes (200 to 400 km) at rather low inclinations the dose rate is quite low even without protection.

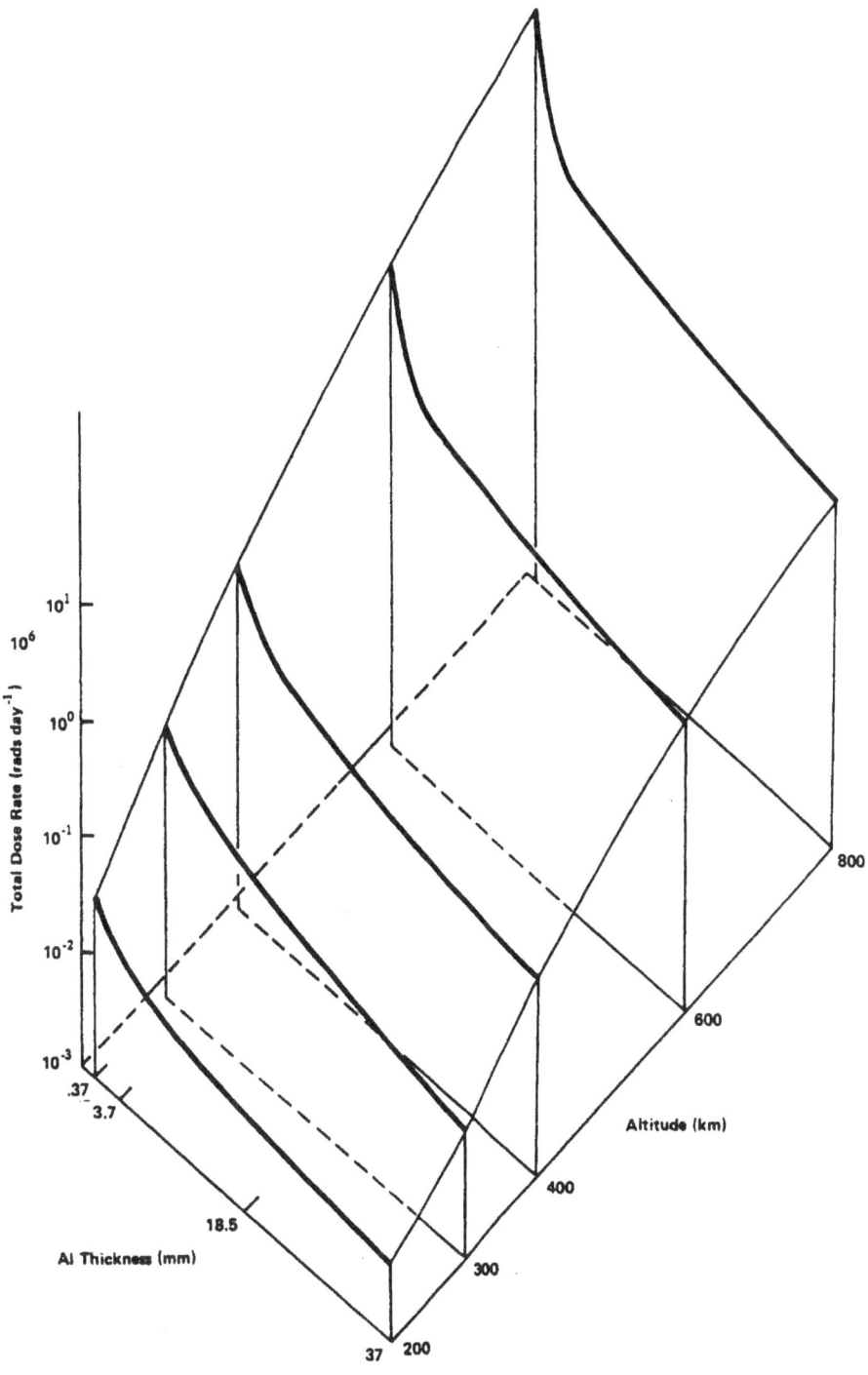

Figure 4.3-11 Total Dose Rate behind a Spherical Shell at 28.5° Inclinations

Table 4.3-3
*Gyroradii and Periods of the Motions of Particles
in the Guiding Center Approximation* *

L = 1.5				
	R_c (cm)	T_1 (sec)	T_2 (sec)	T_3 (min)
Electrons				
50 keV	4.7×10^3	4.2×10^{-6}	.30	710
500 keV	1.8×10^4	7.7×10^{-6}	.14	90
5 MeV	1.1×10^5	4.2×10^{-5}	.12	12
Protons				
100 keV	2.8×10^5	7.1×10^{-3}	8.4	340
1 MeV	8.8×10^5	7.1×10^{-3}	2.7	34
10 MeV	2.8×10^6	7.2×10^{-3}	.85	3.4
500 MeV	2.2×10^7	1.1×10^{-2}	.16	.082
L = 4.0				
	R_c (cm)	T_1 (sec)	T_2 (sec)	T_3 (min)
Electrons				
50 keV	9.0×10^4	8.1×10^{-5}	.79	270
500 keV	3.4×10^5	1.5×10^{-4}	.38	34
5 MeV	2.1×10^6	7.9×10^{-4}	.33	4.6
Protons				
100 keV	5.3×10^6	1.3×10^{-1}	22.0	130
1 MeV	1.7×10^7	1.3×10^{-1}	7.1	13
10 MeV	5.3×10^7	1.4×10^{-1}	2.3	1.3
500 MeV	—	—	—	—

*The gyroradii and periods have been computed according to the formulas of Hamlin, Karplus, Vik, and Watson. J. Geophys. Res., 66.1-4 (1961). A dipole field is assumed. The periods are calculated for particles which mirror at a geomagnetic latitude of 30. The gyroradius is given for the instant at which the particle crosses the geomagnetic equator.

4.4 MAGNETIC AND ELECTRIC FIELDS

4.4.1 Magnetic Fields

The geomagnetic field is characterized (at any point) by its direction and magnitude. These are specified by two direction angles and the magnitude by its three perpendicular magnitudes. The magnitude can be given in Oersted (magnetic intensity) or a gauss (magnetic induction). The field is less than one oersted so that the gamma unit is often used: one gamma equals 10^{-5} oersted or 10^{-5} gauss.

Some of the angles (D and I) and the components (X, Y, Z) commonly used are shown in Figure 4.4-1. The vector geomagnetic field $\overline{F}$ has the magnitude F, the total intensity or total field. The horizontal vector component $\overline{H}$ has the magnitude H, the horizontal intensity. The vertical vector component $\overline{Z}$ has the vertical magnitude Z. The northward, eastward, and downward components X, Y, Z are the Cartesian components of the field. The magnitude of the angle between $\overline{H}$ and $\overline{X}$ is the declination D. The magnitude of the angle between $\overline{H}$ and $\overline{F}$ is the inclination or dip.

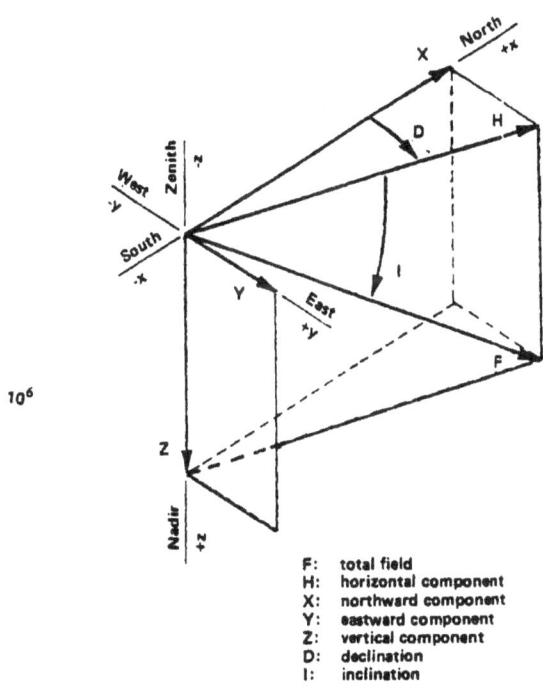

F: total field
H: horizontal component
X: northward component
Y: eastward component
Z: vertical component
D: declination
I: inclination

Figure 4.4-1 Definition and Sign Convention for the Magnetic Elements

The field has been varying drastically over geological time. The portion which varies with a period greater than about a year is considered as the steady field while the remainder is considered as the variation field.

Most of the steady field arises from sources below the surface of earth (excluding induced currents in the earth by external sources) and is known as the main field. About ten percent of the main field consists of large-scale anomalies (up to thousands of kilometers) and small-scale irregularities (on the order of 10 kilometers). These are termed the residual field. The slow change in the main field, with time constants of tens to thousands of years, is called the secular variation.

The variation fields consist of the quiet variation and the disturbed variation field. The quiet fields result from periodic variations in gravity force, solar illumination, and compression or other modification by solar-wind effects. These vary diurnally and seasonally. The so-called Sq (solar quiet) variation field results from solar electromagnetic radiation which heats and ionizes the atmosphere which in turn produces convective flow and high conductivity in the ionosphere. The motion of a conducting fluid in the presence of the mainfield generates currents which produce the Sq field. At most, the peak-to-peak amplitudes of this field are of several tens of gammas at the surface. The tidal flow of the atmosphere from the luni-solar gravitation field generates the so-called L (lunar daily) field with an amplitude at the surface of about a tenth of that of the Sq field. Another contribution rises from the confinement of the main field by the solar wind. This compression is stronger in the daytime leading to a diurnal variation of a few gammas at the surface. In more distant regions of the magnetosphere, it is dominant, completely altering the field configuration.

The disturbed variation fields do not have a simple periodicity and seem to result from changes in the interplanetary environment. They are also termed the geomagnetic disturbance or the D field and are those fields which remain after the steady and quiet variation fields have been subtracted from the total. Details of these fields are discussed in Reference I-21.

An approximation to the geomagnetic field near the Earth's surface (up to about 2000 km) is an Earth centered dipole with its axis tilted to 78.5°N, 291.0°E for the geomagnetic north pole and its geomagnetic south pole at 78.5°S, 110°E. In spherical coordinates r, θ and ϕ, the magnetic scalar potential in a spherical-harmonic expansion is (Ref. I-22)

$$V = R_e \sum_{n=1}^{\infty} \sum_{m=0}^{n} P_n^m (\cos\theta) [(\frac{R_e}{r})^{n+1} (g_n^m \cos m\phi + h_n^m \sin m\phi) + (\frac{R_e}{r})^{-n} (A_n^m \cos m\phi + B_n^m \sin m\phi)] \quad 4.3\text{-}9$$

where r, θ, and ϕ are the geographical polar coordinates of radial distance, colatitude (dipole axis), and east latitude, and R_e is the radius of the Earth. The functions $P_n^m(\cos\theta)$ are the Schmidt functions

$$P_n^m(\cos\theta) = [\frac{\epsilon_m (n-m)!}{(n+1)!}]^{1/2} [\frac{(1-\cos^2\theta)^{m/2}}{2^n n!}][\frac{d^{m+n}}{d(\cos\theta)^{n+m}}](\cos^2\theta - 1)^n \quad 4.3\text{-}10$$

$\epsilon_m = 2$ if $m > 0$

$\epsilon_m = 1$ if $m = 0$

The mean-square value of $P_n^m(\cos\theta)$ integrated over the sphere is $(2n+1)-1/2$.

In the potential, the terms containing g_n^m and h_n^m are from sources internal to the Earth and the terms containing A_n^m and B_n^m arise from external currents. The potential function is valid in the space above the surface and below the external current rupture. The field then is

$$\overline{B} = -\nabla V \qquad 4.3-11$$

with northward, eastward, and downward components

$$X = \frac{1}{r}\frac{\partial V}{\partial \theta},$$

$$Y = \frac{1}{r\sin\theta}\frac{\partial V}{\partial \phi}, \text{ and}$$

$$Z = \frac{\partial V}{\partial r}.$$

Discussion, values of Schmidt coefficients, and computer programs to calculate the field are given in References I-23 through I-26. Figure 4.4-2 shows contours of constant field intensity F for the year 1965.

4.4.2 Electric Field

In the presence of the magnetic field, the electric field that is observed depends on the frame of reference in which it is measured. Let $\overline{E_1}$ and $\overline{B_1}$ be the fields in frame 1, then in a second frame, frame 2, moving with a velocity $v \ll c$ (c is the velocity of light) relative to frame 1, then (Ref. I-16)

$$\overline{E_2}_\perp = \overline{E_1}_\perp + (v/c)\overline{B} \qquad \text{(cgs units)} \qquad 4.3-14$$

where $\overline{B_2} = \overline{B_1}$,

$\overline{E_2}_\parallel = \overline{E_1}_\parallel$, and

where $\perp$ and $\parallel$ are the perpendicular and parallel components, respectively, measured with respect to $\overline{B}$. In more useful limits

$$E_{2\perp}(\text{volts/m}) = E_{1\perp}(\text{volts/m}) + \frac{v(\text{m/s})\cdot \overline{B}(\text{gauss})}{10^4}$$

In the lower ionosphere where the plasma is partly tied to the neutrals by high collision rates an electric field may exist in the neutral frame. The current, $\overline{J}$, is given by (Ref. I-25)

$$\overline{J} = \sigma_p \overline{E}_\perp - \sigma_h \frac{\overline{E}\times\overline{B}}{B} + \sigma_\parallel \overline{E}_\parallel \qquad 4.3-15$$

where

σ_p = Pedersen conductivity,
σ_h = Hall conductivity, and
$\sigma_\parallel$ = intrinsic or parallel conductivity.

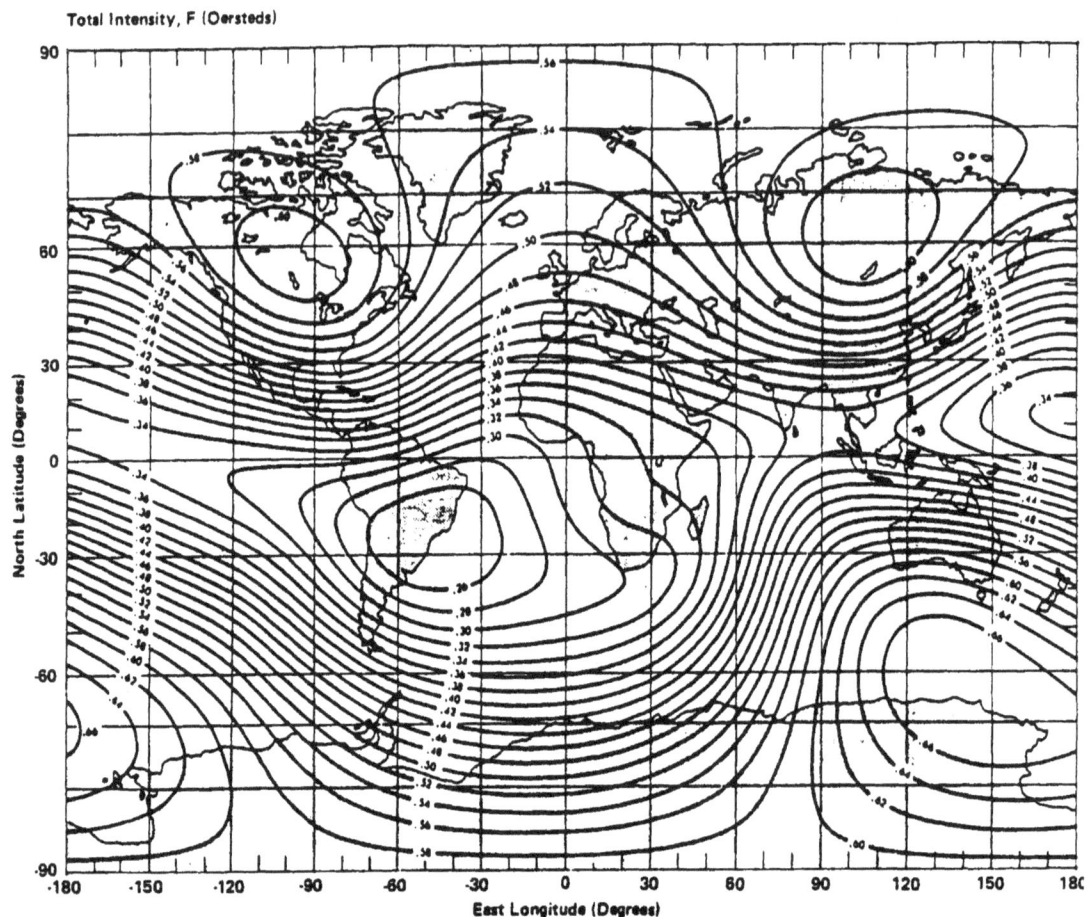

Figure 4.4-2 Contours of Constant F (Total Field) for IGRF 1965.0

Figure 4.4-3 shows typical conductivities as a function of altitude. There are various analytical models given in Reference I-26 which give both the spatial and temporal dependence.

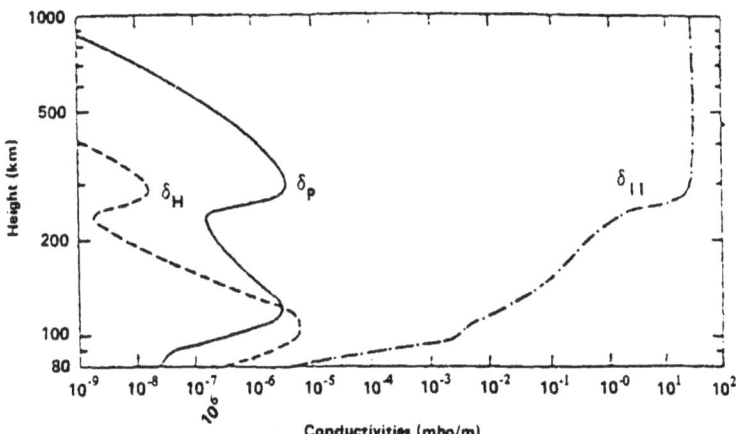

Figure 4.4-3
Typical Variation of the Ionospheric Conductivities with Height for a Nighttime Ionosphere (Ref I-15)

4.5 OTHER ENVIRONMENT REFERENCES

I-1. Filz, R. C., L. Katz, G. A. Kuck, M. A. Shea, and D. F. Smat, "Corpuscular Radiation", Air Force Cambridge Research Laboratories, Dec. 1968.

I-2. Minzner, R. A., ed., "The 1976 Standard Atmosphere above 86 km Altitude", NASA SP-398, 1976.

I-3. Jacchia, L. G., "Revised Static Models of the Thermosphere and Exosphere with Empirical Temperature Profiles", Smithsonian Astrophys. Obs. Spec. Rep. No. 332, Cambridge, Mass., 1971.

I-4. Jacchia, L. G., "Thermospheric Temperature, Density and Composition: New Models," Smithsonian Astrophys. Obs. Spec. Rep. No. 375, Cambridge, Mass., 1977.

I-5. Johnson, D. L. and R. E. Smith, "The MSFC/J70 Orbital Atmosphere Model and the Data Bases for the MSFC Solar Activity Prediction Technique", NASA TM-86522, November 1985.

I-6. Johnson, D. L., "Global Matrix of Thermospheric Density Values for Selected Solar/Geomagnetic Conditions and Spacecraft Orbital Altitudes", NASA TM-86478, December 1984.

I-7. Hedin, A. E., C. A. Reber, G. P. Newton, N. W. Spencer, H. C. Brinton, H. C. Mayr, and W. E. Potter, "A Global Thermospheric Model Based on Mass Spectrometer and Incoherent Scatter Data MSIS2. Composition", J. Geophys. Res., 82, 2148, 1977.

I-8. Hedin, A. E., "Tables of Thermospheric Temperature, Density, and Composition Derived from Satellite and Ground Based Measurements", NASA-GSFC, January 1976.

I-9. Hickman, D. R., B. K. Ching, C. J. Rice, L. R. Sharp and J. M. Straws, Space Sciences Lab., The Aerospace Corporation, 1978.

I-10. "The Earth's Ionosphere", NASA SP-8049, March 1971.

I-11. Langton, H. N., ed., "The Space Environment", Elsevier Pub Co.

I-12. Valley, S. L., ed., "Handbook of Geophysics and Space Environments", U. S. Air Force, 1965.

I-13. MacIntosh, P. S. and M. Dryer, eds., "Solar Activity and Predictions", MIT Press, Cambridge Mass, 1972.

I-14. Stassinopoulos, E. G. and J. H. King, "IEEE Trans. Aerospace and Electronic Sys.", $\underline{10(4)}$, 442, 1974.

I-15. "Space and Planetary Environment Criteria Guidelines for Use in Space Vehicle Development", NASA TM-78119, 1977.

I-16. "Space and Planetary Environment Criteria Guidelines for Use in Space Vehicle Development", 1982 Revision NASA TM-82478, 1983.

I-17. Diedrich, Martin Marietta Internal Report, July 1978.

I-18. Teague, M. J. and J. I. Vette, "The Inner Zone Electron Model, AE-6" NSSDC 76-4, NASA-GSFC, 1976.

I-19. Sawyer, D. M. and J. I. Vette, "AP-8 Trapped Proton Environment for Solar Maximum and Minimum", NSSDC 76-06, NASA-GSFC, 1976.

I-20. Fairbridge, R. W., ed., "Encyclopedia of Atmospheric Sciences and Astronomy", Reinholt Pub. Corp., New York, 1967.

I-21. Watts, J. W., Jr. and J. J. Wright, "Charged Particle Radiation Environment for the Spacelab and Other Missions in Low Earth Orbit", Revision A, NASA TMX-73358, 1976.

I-22. Knecht, D. J., "The Geomagnetic Field", Air Force Cambridge Research Laboratories, 1972.

I-23. Peddle, N. W., "International Geomagnetic Reference Field 1980. A report by IAGA Division 1 Working Group 1", Geophys. J. R. Astr. Soc., vol. 68, pp. 265-268, 1982.

I-24. "IAGA Division 1, Study Group; International Geomagnetic Reference Field 1975", J. Geophys. Res., vol. 81, pp. 5163-5164, 1977.

I-25. "Magnetic Fields Earth and Extraterrestrial", NASA SP-8017, March 1969.

I-26. Zmuda, A. J., ed., "World Magnetic Survey 1957-1969", Bulletin 28, Int. Assn. Geomagn. and Aeron., 1971.

I-27. Olson, W. P., ed., "Quantitative Modeling of Magnetospheric Processes", Am. Geophys. Union, Washington, D.C., Geophysical Monograph 21, 1979.

Appendix A—Materials Properties

APPENDIX A

SPACECRAFT MATERIALS PROPERTIES

It is beyond the scope of this or any handbook to present a complete list of properties for all the materials an experimenter might like to use in constructing a Spacelab experiment. Instead, references where properties may be located will be presented.

In the area of outgassing, Reference J-1 presents a standardized test technique for determining the total vacuum weight loss and condensible percentages in a 24 hour period. Reference J-2 presents test results for thousands of materials which have been tested. To be approved, a material must have less than 1 percent TWL and less than 0.1 percent VCM.

For oxidation and glow, the most complete compilation of properties data is contained in Section 3 of this Handbook.

For optical properties, Reference J-3 presents properties for several thousand materials. Reference J-4 presents properties for materials with specific aerospace applications

Materials Properties References

J-1. Leger, L. J., "General Specification Vacuum Stability Requirements of Polymeric Materials for Spacecraft Application", SP-R-0022A, NASA JSC, Houston, Texas, 9 September 1974.

J-2. "Compilation of VCM Data of Nonmetallic Materials", JSC 08962, Rev. U, with Addenda, MDTSCO, Houston, Texas.

J-3. Touloukian, Y. S. (ed.), and C. Y. Ho (tech. ed.), "Thermophysical Properties of Matter", Vol. 1-13, Thermophysical Properties Research Center, Purdue University.

J-4. Touloukian, Y. S., and C. Y. Ho (eds.), "Thermophysical Properties of Selected Aerospace Materials, Part 1, Thermal Radiative Properties", Thermophysical Properties Research Center, Purdue University, 1976.

www.ingramcontent.com/pod-product-compliance
Lightning Source LLC
Chambersburg PA
CBHW081721170526
45167CB00009B/3659